Digital Circuits

JOHN CANAVAN

L_EROY STEVENS
Phoenix College

RINEHART PRESS
SAN FRANCISCO

© 1973 by Rinehart Press
5643 Paradise Drive
Corte Madera, Calif. 94925

A division of Holt, Rinehart and Winston, Inc.

All rights reserved.

Library of Congress Catalog Card Number: 79-136169

ISBN: 0-03-084881-4

PRINTED IN THE UNITED STATES OF AMERICA

3 4 5 6   090   9 8 7 6 5 4 3 2 1

Preface

This book is intended to be used in technical institutes, community colleges, four-year institutions, or any educational institution that offers a program in electronics technology. Several good texts have been written on this subject; however, many of these books are not designed to be completed in one semester. This text was written from the viewpoint of teachability and flexibility.

The content is structured so that it may be used in a short term course, with or without the emphasis on mathematics. This flexibility is possible because a mathematical treatment has been employed, but the analyses are sufficiently simple to permit the reader to grasp the fundamental concepts. One only needs to have a knowledge of algebra and basic trigonometry to work problems. If the instructor chooses not to emphasize mathematical solutions to reinforce basic principles, there are ample review questions at the end of each chapter to afford the reader additional profitable information. Graphical solutions of problems are emphasized where applicable throughout the book.

As a prerequisite, the reader should have an understanding of ac and dc circuit theory, including network theorems, and a knowledge of active devices (vacuum tubes and transistors), at least at the introductory level. About 95% of the book is concerned with solid state electronics. The justification for including any vacuum tube application at all is that in industry there still exists a considerable amount of vacuum tube test equipment. Technicians will be called upon for several years to come to repair and modify these instruments. Furthermore, the Federal Communication Commission tests and the broadcast industry in general still require the use of vacuum tube technology.

Grateful acknowledgment is made to the staff of the Electronics Department at Phoenix College for their encouragement and advice during the final preparation of the manuscript.

JOHN CANAVAN
LEROY STEVENS

Contents

1 TRANSIENTS AND LINEAR WAVESHAPING — 1

    1-1 The *RC* Circuit — 2
    1-2 The *RL* Circuit — 3
    1-3 The Time Constant — 4
    1-4 The Differentiating Circuit — 7
    1-5 The *RL* Circuit as a Differentiator — 9
    1-6 The Differentiator as a High-Pass Filter — 9
    1-7 The Integrating Circuit — 10
    1-8 The *RL* Integrator — 12
    1-9 The Integrator as a Low-Pass Filter — 13

2 DIODES AND DIODE APPLICATION — 16

    2-1 The Semiconductor Diode — 16
    2-2 The Diode Equation — 18
    2-3 The Characteristic Curve for the Diode — 18
    2-4 Diode Response Time — 21
    2-5 Diode Clipping Circuits — 22
    2-6 Diode Clamps — 25
    2-7 Clamping the Nonsymmetrical Signal — 29
    2-8 Ground and Delay Clamping Circuit — 31

3 TRANSISTOR AND VACUUM TUBES — 35

    3-1 The Junction Transistor — 35
    3-2 Characteristic Curves for the Transistor — 38
    3-3 Leakage Currents — 39

	3-4	The Hybrid or h Parameters	40
	3-5	The Transistor as a Switch	42
	3-6	Response of the Transistor Switch	45
	3-7	Vacuum Tubes	48
	3-8	The Triode as a Switch	48
	3-9	Pentodes	51

4 LOGIC GATES AND TRANSMISSION GATES — 54

4-1	Logic	54
4-2	The AND Gate	57
4-3	The OR Gate	58
4-4	The NOT Circuit	59
4-5	Other Forms of Logic Circuits	62
4-6	Transmission Gates	62
4-7	The Unidirectional Diode Gate	63
4-8	The Bidirectional Gate	64
4-9	Elimination of the Pedestal	66
4-10	A Bidirectional Diode Gate	66
4-11	Nonideal Conditions	68

5 MULTIVIBRATORS — 70

5-1	The Bistable Multivibrator (Flip Flop)	70
5-2	Triggering the Flip Flop	72
5-3	Saturating vs Nonsaturating Flip Flops	74
5-4	Cathode Coupled Binary (Schmitt Trigger)	75
5-5	The Monostable Multivibrator (One Shot)	78
5-6	The Vacuum-Tube One Shot	81
5-7	Emitter Coupled One Shot	82
5-8	The Astable Multivibrator	85

6 THE BLOCKING OSCILLATOR — 90

6-1	The Ideal Pulse Transformer	90
6-2	The Nonideal Pulse Transformer	94
6-3	The Monostable Blocking Oscillator	96
6-4	Triggering the Blocking Oscillator	99
6-5	The Free-Running (Astable) Blocking Oscillator	99
6-6	Synchronization of the Astable Blocking Oscillator	102

7 TIME BASE GENERATORS 106

  7-1 Sweep Circuit Linearity 107
  7-2 The Gas Tube Sweep Circuit 110
  7-3 The Thyratron Sweep Circuit 112
  7-4 Synchronization of the Thyratron Sweep Generator 114
  7-5 The Triode Integrator 115
  7-6 The Transistor Integrator 116
  7-7 The Miller Integrator 117
  7-8 Miller Sweep Circuit with a Pentode 121
  7-9 The Phantastron Circuit 122
  7-10 The Bootstrap Sweep Circuit 123

8 NEGATIVE RESISTANCE DEVICES 128

  8-1 The Tunnel Diode 129
  8-2 The Tunnel Diode as a Switch 131
  8-3 The Tunnel Diode One-Shot Circuit 131
  8-4 Linearization of the Characteristic Curve 134
  8-5 A Bistable Tunnel Diode Circuit 136
  8-6 The Astable Tunnel Diode Circuit 137
  8-7 Tunnel Diode Hybrid Circuits 139
  8-8 The Unijunction Transistor 141
  8-9 A Unijunction Relaxation Oscillator 145
  8-10 Negative Resistance Devices in General 148

  INDEX 151

Transients and Linear Waveshaping

1

The primary objective of this chapter is to deal with circuits that have as their driving source a function other than a sinusoidal waveform. Waveforms that represent different (from sinusoidal) basic mathematical functions are important because they specify the natural impulses and responses encountered in pulse circuitry. Each of the basic functions is illustrated in Fig. 1-1 with the equation that describes the particular function. It will become apparent later in the chapter that periodic repetitions of each of these functions are also commonplace. The purpose of this section is to explore the generation of some functions and the response of certain linear circuits to them. Discussion begins with the review of transients in RC and RL circuits, followed by

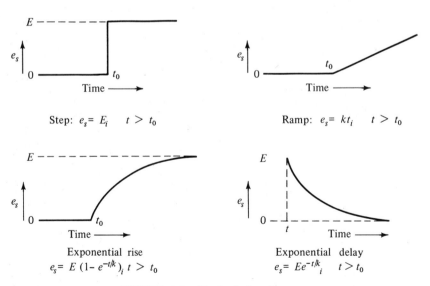

FIGURE 1-1. The basic functions.

a more detailed discussion of the fundamental differentiating and integrating networks. The level of mathematics incorporated in the chapter involves algebra and solution of exponential equations. An understanding of calculus would be helpful but is not necessary to understand the theory and to solve problems. Calculus is used to enlighten the student as to the origin of basic equations discussed in the book.

1-1 THE *RC* CIRCUIT

Consider a series circuit consisting of a resistor, capacitor, battery, and switch as shown in Fig. 1-2. Assuming that the capacitor has no initial charge, the circuit operation begins with the time when the switch is placed in position 1. The circuit equation is

$$E = e_C + e_R \tag{1-1}$$

$$= \frac{q}{C} + R\frac{dq}{dt}. \tag{1-1a}$$

Equations of this type yield the following solution:

$$q = Qe^{pt} \tag{1-2}$$

$$\frac{Q}{C}e^{pt} + RQpe^{pt} = 0, \tag{1-3}$$

from which

$$\frac{1}{C} + pR = 0 \tag{1-4}$$

$$p = -\frac{1}{RC}. \tag{1-4a}$$

After a long period of time has elapsed, current will no longer flow; therefore, $e_R = 0$. Then

$$E = e_C = \frac{Q}{C}. \tag{1-5}$$

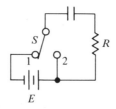

FIGURE 1-2. *RC* charging circuit.

The *RL* Circuit

Since $Q = CE$, the solution for the transient component of the charge is

$$q = CEe^{-t/RC} \tag{1-6}$$

and the current in the circuit

$$i = \frac{dq}{dt} = -\frac{E}{R}e^{-t/RC}. \tag{1-7}$$

Thus, at any time t after the voltage E is applied, Eq. (1-1) can be written

$$E = e_C + Ee^{-t/RC} \tag{1-8}$$

or

$$e_C = E(1 - e^{-t/RC}) \tag{1-9}$$

$$e_R = Ee^{-t/RC}. \tag{1-10}$$

Now suppose that after some time t_1, the switch is thrown from position 1 to position 2. The capacitor will have charged to some voltage E_0 prior to t_1. At the first instant the switch is in position 2, the capacitor will appear as a voltage source E_0 in series with the resistor. Note, however, that the polarity of this voltage source is such that current must flow in the opposite direction as it did when the switch was in position 1. Now the complete circuit equation is

$$e_C + e_R = 0. \tag{1-11}$$

Thus, at any time t_2 after t_1,

$$e_C = E_0 e^{-(t_2-t_1)/RC} \tag{1-12}$$

$$e_R = -e_C. \tag{1-13}$$

Note: Instead of solving for current one can also state $e_C = E - (E - E_0)e^{-t/RC}$, where E_0 represents initial voltage on C and E the applied voltage.

$$e_C = E - (E - E_0)e^{-t/RC}. \tag{1-14}$$

1-2 THE *RL* CIRCUIT

If the capacitor in the foregoing analysis is replaced with an inductor, a similar analysis follows. In fact, the solution is more direct due to the fact that the time varying quantity is current (instead of charge) in the basic equation. For the circuit of Fig. 1-3 when the switch is placed in position 1,

$$E = L\frac{di}{dt} + Ri. \tag{1-15}$$

At any time t after the switch is placed in position 1,

$$i = \frac{E}{R} - \frac{E}{R}e^{-(R/L)t}. \tag{1-16}$$

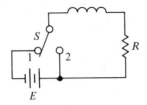

FIGURE 1-3. RL charging circuit.

E/R is the steady state current; that is, if the switch remains in position 1 a long period of time, the current will reach a maximum value equal to E/R. Therefore,

$$I_s = \frac{E}{R}. \tag{1-17}$$

Had there been an initial current I_o flowing in the circuit prior to t_0, then

$$i = \frac{E}{R} - \left(\frac{E}{R} - I_o\right) e^{-(R/L)t} \tag{1-18}$$

$$i = I_s - (I_s - I_o)e^{-(R/L)t} \tag{1-19}$$

equations that are applicable for either position of the switch in Fig. 1-2. Thus it follows that when the switch is placed in position 2 after having been in position 1 for some time

$$i = Ie^{-(R/L)t}. \tag{1-20}$$

In Eq. (1-20) I is the current value reached while the switch is in position 1, and t is the time after the switch is placed in position 2.

1-3 THE TIME CONSTANT

The exponential quantity in the equations for transient circuit analysis is the determining factor for the solution of the equations. No solution is available until a value of t is specified in addition; however, the coefficient of t is a circuit parameter that determines the extent to which time is significant. In other words, the length of time the transient effect need be of concern after a disturbance is determined by the value of the coefficient of t. The reciprocal of this coefficient is called the time constant (τ) for the circuit. Thus, the time constant is L/R or RC for the RL and RC circuits, respectively. By measuring time in increments whereby the basic unit is the time constant, the expression for $e^{-t/\tau}$ can be plotted and applied in the solution of any simple

The Time Constant

circuit. (Of equal value in the solution of transient problems is the graph of the function $1 - e^{-t/\tau}$.) The *RC* time constant, simply stated, is: the amount of time t required for the voltage across the capacitor to rise to 63 percent of the applied voltage E. Also, the L/R time constant may be stated as the amount of time t required for the current I to rise to a value of 63 percent of the steady state current. The proof of these statements is demonstrated by the graph of Fig. 1-4. Notice that when t/τ is 1, the percent of change for either curve is 63 percent.

Example 1-1 The *RC* circuit of Fig. 1-2 has the values of 10 kΩ, 50 μF, and 10 V for *R*, *C*, and *E*, respectively.
(a) Determine the time constant.
(b) Calculate the voltage across the resistor 750 msec after the switch is placed in position 1.

Solution:
(a)
$$\tau = RC$$
$$= (10 \times 10^{+3})(50 \times 10^{-6})$$
$$= .5 \text{ sec.}$$

(b) Assuming the switch had been in position 2 for a long time, there would be no initial charge on *C*. Let $t = 750 \times 10^{-3}$ sec. Then $t/\tau = 1.5$, and from Fig. 1-4, $e^{-t/\tau} = .23$.

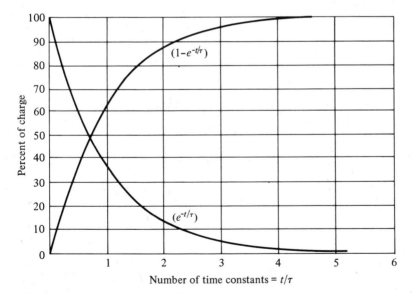

FIGURE 1-4. The universal time constant graph.

From Eq. (1-10),
$$e_R = Ee^{-t/\tau}$$
$$= 10(.23)$$
$$= 2.3 \text{ V}.$$

The answer arrived at for (b) of Example 1-1 could just as well have been obtained by calculation. However, the graph of Fig. 1-4 permits one to read the value for $e^{-1.5}$ directly.

Of more importance is the fact that, once the value for τ is known, one can mentally determine the duration of the transient period for the circuit. A commonly used standard is to assume that the transient period ends after five time constants have elapsed.

Example 1-2 Consider the RC circuit of Fig. 1-5 and note that the switch has been open for a long time. The switch is then placed in position 1 for .2 sec, after which it is placed in position 2 and left there. Sketch the waveform for e_o.

Solution: The circuit response will be affected by two time constants, the charging and discharging time constants. The capacitor will charge for .2 sec when the switch is placed in position 1. The expression for current when the switch is in position 1 is

$$i = Ie^{-t/RC},$$

where

$$I = \frac{E}{R} = \frac{50 \text{ V}}{1 \text{ k}\Omega} = 50 \text{ mA}.$$

Then

$$e_o = iR = 50e^{-t/.1 \text{ sec}}$$

is the expression for e_o at any time after the switch is placed in position 1 and before it is placed in position 2.

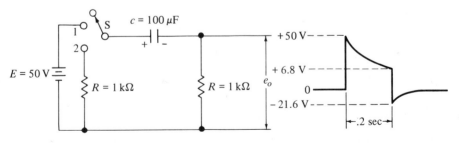

FIGURE 1-5. RC network.

The Differentiating Circuit

After two time constants, the charging current will have decayed 86.4 percent (the capacitor will have charged to 86.4 percent of its final value), as e^{-2} will be .136 and e_o will be $+6.8$ V.

When the switch is placed in position 2, the voltage accumulated on C (43.2 V) will decay, causing current to flow (opposite to the direction of flow in the previous part of the problem) through the two resistors. The time constant during the time is

$$\tau = RC = 200 \text{ msec}$$

and

$$i = I_0 e^{-t/.2}.$$

At the instant of switching ($t_2 = 0$),

$$i = I_0 = \frac{e_C}{R} = \frac{43.2 \text{ V}}{2 \text{ k}\Omega} = 21.6 \text{ mA}$$

$$e_o = -iR = -21.6 \text{ V},$$

from which value e_o will decay toward 0. Notice that the instant the switch is placed in position 2 the capacitor voltage is 43.2 V, which is divided equally between the two 1 kΩ resistors. Thus, e_o is -21.6 V.

The foregoing example points out the important fact that the output voltage can become negative, depending upon the reference point, even though the externally applied voltage was positive.

One last comment is in order. The transient circuits discussed have been single-energy storage circuits. Double-energy storage circuits can be handled in the preceding manner only when the time constants associated with each storage element are very different. This, in effect, permits the transient effect in one part of a circuit to precede the transient action in another part of the circuit. Where doubt exists, the best approach is to apply Kirchhoff's rules, which will result in differential equations (simultaneous for more than one loop).

1-4 THE DIFFERENTIATING CIRCUIT

Consider the RC circuit of Fig. 1-6.

$$e_i = e_C + e_R = \frac{1}{C} \int i \, dt + iR$$

$$i = C \frac{de_C}{dt}. \tag{1-21}$$

If, by the proper choice of components, e_R is made small, then

$$e_i \approx e_C$$

Transients and Linear Waveshaping

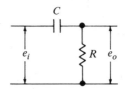

FIGURE 1-6. The RC differentiator.

or

$$i \approx C \frac{de_i}{dt} \qquad (1\text{-}22)$$

$$e_o = iR \approx RC \frac{de_i}{dt}. \qquad (1\text{-}23)$$

Thus, the output voltage is nearly equal to a constant times the rate of change of the input voltage with respect to time (de_i/dt). This, in essence, is the definition of a differentiating circuit. In practice, it is not difficult to make the approximation quite good. It is only required that the RC time constant be made small with respect to the period of the applied voltage.

A square wave may be considered as a periodic repetition of two constant voltages as shown in Fig. 1-7. During the times t_1 and t_2 the input voltage is constant; therefore, according to previous theory, the output voltage should be zero. That is because no change of input voltage occurred with respect to the time interval ($de_i/dt = 0$). However, the two constant voltages of the square wave are separated by very brief discontinuities. Recall from basic theory that an inherent characteristic of a capacitor is that the voltage across

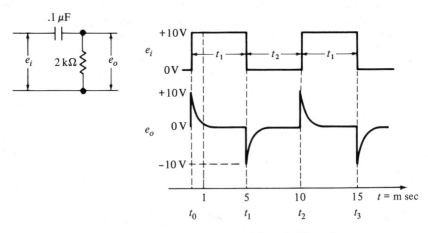

FIGURE 1-7. Input-output response of the RC differentiator.

the capacitor cannot be changed instantaneously. Therefore, in these short periods of time the RC circuit will have a transient response.

Example 1-3 Sketch the output waveform for the applied input square wave of Fig. 1-7.

Solution: At t_0 the input rises from 0 to 10 V. We assume this to be an instantaneous change. The transient period can be considered complete after five time constants; therefore, in

$$5\tau = 1 \text{ msec}$$

the capacitor will have changed to 10 V, no current will be flowing in the circuit, and e_o will be 0.

At t_1 the input voltage will drop to 0 instantaneously (the same as if a switch had shorted the input), and the capacitor will discharge. Since the capacitor is incapable of supporting an instantaneous change in voltage, the total initial voltage change will be impressed across R. Five time constants later, the transient will have decayed and the output risen to 0.

The sketch exaggerates the transient period with respect to the overall time t_1 and t_2. In actuality, very sharp spikes or peaks can be produced. Note that except at the discontinuity, the output voltage is zero, as should be expected from the derivative of a constant.

1-5 THE RL CIRCUIT AS A DIFFERENTIATOR

The RL circuit can also be used as a differentiator; again, with the requirements for a short time constant. The RC circuit is generally more desirable, however, because of the resistance drop in the inductor.

1-6 THE DIFFERENTIATOR AS A HIGH-PASS FILTER

If a sine wave is applied to the circuit of Fig. 1-6, the output voltage is a function of the relative value of the capacitive reactance to the resistance. When the capacitive reactance is equal to the resistance, the output voltage is .707 of e_i, the input voltage. This drop in output level corresponds to an output reduction of 3 dB, and the frequency that this condition exists is referred to as the lower cutoff frequency f_1. The output voltage will be a sine wave shifted by an angle θ such that

$$\tan \theta = \frac{X_C}{R} = \frac{1}{2\pi f_1 RC} \qquad \text{(1-24)}$$

if
$$X_C = R, \quad \theta = 45°.$$
Then
$$\tan 45° = 1.$$
Therefore,
$$f_1 = \frac{1}{2\pi RC}. \tag{1-25}$$

The cutoff frequency of a high-pass filter is clearly dependent on the time constant τ of the circuit. With the proper choice of component values, differentiation can be nearly ideal, and it is not unreasonable to expect a cosine wave output where a sine wave is applied.

Example 1-4 Determine the cutoff frequency if the circuit of Fig. 1-6 has the value of 5 kΩ and .006 μF for R and C, respectively. The input signal is a sine wave driving function.

Solution:
$$f_1 = \frac{1}{2\pi RC} = 5.3 \text{ kHz.}$$

A question should be asked at this point. Is this circuit and choice of component values an ideal differentiator circuit? The answer is no, because there must be an approximately 90° phase shift of e_o to e_i to obtain a cosine curve; the phase shift in this circuit is only 45°.

1-7 THE INTEGRATING CIRCUIT

The integrating circuit parallels the differentiator in its mathematical approach. The circuit is shown in Fig. 1-8. Again, the circuit equation is

$$e_i = e_C + e_R = \frac{1}{C}\int i\,dt + iR. \tag{1-21}$$

If e_C is small, $e_i \approx e_R$, or

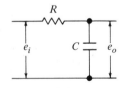

FIGURE 1-8. The RC integrator.

The Integrating Circuit

$$i \approx \frac{e_i}{R}$$

$$e_o = e_C \approx \frac{1}{RC} \int e_i \, dt. \qquad (1\text{-}26)$$

As with the differentiator, the circuit works quite well in practical applications. The important restriction with the integrator is that the RC time constant be large with respect to the period of the applied voltage.

The integrating circuit may be defined as a circuit whose output voltage is the integral of the input voltage $\int e_i \, dt$. This means that the voltage out will be the final value of all the instantaneous values of the input voltage added and averaged over a particular interval of time, multiplied by a constant.

Before investigating an actual problem, a closer analysis of Eq. (1-9) and the graph of Fig. 1-4 is needed. Observe that if t/τ is small or τ is large, the percent of change is quite linear, and the output voltage e_C or e_R will also be linear. This means that the capacitor never begins to fully charge but charges only during the linear part or start of the transient time. The following example should be studied while keeping this point in mind.

Example 1-5 Sketch the output waveform for the applied input square wave of Fig. 1-9. Assume that $t_1 = t_2 = 500$ μsec.

Solution: Since $e_C = e_o$,

$$e_o = E - (E - E_0)e^{-t/RC},$$

where

$E =$ applied voltage
$E_0 =$ voltage on C at t_0.

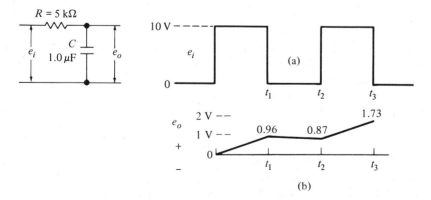

FIGURE 1-9. Input-output response of the RC integrator.

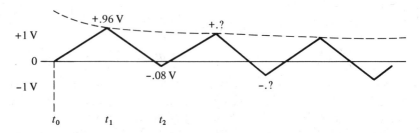

FIGURE 1-10. Integrated output response.

Initially,
$$E_0 = 0, \quad E = +10 \text{ V}.$$
Then, after 500 μsec (t_1),
$$e_o = 10 - 10e^{-.1}$$
$$= .96 \text{ V}.$$
Therefore, at t_1, e_o is .96 V, and the input voltage drops from +10 to 0; the circuit has an input voltage of 0 V with an initial e_C of +.96 V.
$$e_o = 0 - (0 - .96)e^{-.1}$$
$$= +.87 \text{ V}.$$
At the beginning of the next 500 μsec period, t_2,
$$e_o = 10 - (10 - .87)e^{-.1}$$
$$= +1.73 \text{ V}.$$

Note that the output voltage becomes more positive each time until eventually a constant level is reached. This waveform is the integration of the input driving function (Fig. 1-9a). One further observation about the previous example is noteworthy. Consider the problem if the input voltage varied from +10 to −10 V, therefore having a peak-to-peak voltage of 20 V. Continuing with the step-by-step analysis as in the example, it becomes apparent that the value e_o goes positive and negative. In this instance the output voltage will eventually reach a constant level where the average value is zero (the area above 0 V will equal the area below) (Fig. 1-10). The proof of this statement by calculation is left as an exercise at the end of the chapter.

1-8 THE *RL* INTEGRATOR

Again, it should be mentioned that integration with the *RL* circuit is possible, and, theoretically, the problem is identical to the one just discussed. In practice, the physical limitations of inductors (that is, internal resistance and capacitance) preclude widespread use.

1-9 THE INTEGRATOR AS A LOW-PASS FILTER

Just as the differentiating circuit is a high-pass filter, so the integrator is a low-pass filter. The only difference between the filter and the integrator is in the choice of components and application. The simple RC phase shifter used in sinusoidal circuits is merely an integration process performed on the input sine wave. If the circuit of Fig. 1-9 has a sine wave input, the upper cutoff frequency (f_2) will be the frequency that is reached when the capacitive reactance equals the resistance R. The achievement of 90° or 180° of phase shift requires one additional stage to approach an ideal integrator.

Review Questions

1. What is generally meant by *linear waveshaping*?
2. What is the meaning of the term *differentiation*?
3. State the definition of the term *RC time constant*.
4. Define the terms f_1 and f_2 frequencies of a network.
5. What is the relative value of the time constant in an *LR* circuit if *R* is very small?
6. An *RC* circuit has a large time constant τ compared to the period of the driving signal. What statement can you make about the change of voltage across the capacitor?
7. What is the meaning of the term *integrating circuit*?
8. Explain the meaning of discontinuities of a waveform.
9. Will increasing the value of *R* in a particular low-pass filter cause the frequency to increase or decrease? Explain your answer.
10. Define the term *L/R time constant*.
11. What are the disadvantages of an *RL* linear waveshaping circuit compared to an *RC* circuit?
12. Does the output waveform of Fig. 1-7 represent the high- or low-frequency components of the input? Explain in detail.
13. Describe the output waveform of the circuit of Fig. 1-7 if *R* is increased considerably.
14. State the relative effect on the output waveform of the circuit of Fig. 1-9 if *C* is made quite small.

Problems

1. A series RC circuit has the values of 100 kΩ and 0.1 μF for R and C, respectively.
   Find: (a) τ
   (b) f_1
   (c) Show by diagram (e_{in} and e_{out}) a high-pass filter connection.
2. A series RL circuit has the values of 10 mH and 100 Ω for L and R, respectively.
   Find: (a) τ
   (b) The frequency of a square wave that would make this circuit a good differentiator.
   (c) Show by diagram (e_{in} and e_{out}) this circuit as a differentiator.
3. Refer to the circuit of Fig. 1-5. The circuit has the values of 100 V, 1 μF, and 100 kΩ for E, C, and R, respectively. Assume that the switch S remains in position 1.
   Find: (a) τ of the circuit
   (b) e_o when t is 200 msec
   (c) e_C when t is 200 msec
   (d) I when t is 200 msec
   (e) f_1
4. Refer to Example 1-5 and sketch the output waveform for two cycles of input. Assume E to be 20 V p-p, or ± 10 V, and $t_1 = t_2 = 500$ μsec.
5. Assume a series RC high-pass circuit has an $f_1 = 100$ Hz and $C = 2$ μF.
   Find: (a) X_C at f_1
   (b) τ
   (c) proper R
6. An RC differentiating circuit has the following circuit parameter. Input voltage is a symmetrical square wave with a period of 1 msec, an average voltage of 4 V, and a peak value of 8 V; $R = 2.5$ kΩ, $C = .01$ μF.
   Find: (a) Sketch the output waveform.
   (b) e_o when t is .025 msec
   (c) I when t is .025 msec
   (d) I when t is zero plus
   (e) I when t is .3 msec
   (f) e_R when t is .04 msec
7. Draw a schematic diagram of an RL integrating circuit. (Show e_{in} and e_{out}.)

Problems

8. Draw a schematic diagram of a two-section L-type low-pass RL filter. (Show e_{in} and e_{out}.)
9. What is the phase shift in a high-pass RC filter if the circuit parameters are $f_1 = 75$ Hz, $C = 0.1$ μF, and $R = 10$ kΩ?
10. Draw a diagram of a two-section L-type RC differentiating circuit.

Diodes and Diode Application

2

This chapter reviews the operation of the semiconductor diode, with an emphasis on those points that impact heavily on the area of pulse circuits. Vacuum-tube diodes will not be discussed because since the advent of semiconductors they have been all but eliminated from the field of low-power, high-speed electronic circuits. Heavy emphasis is placed on the junction diode theory since it is the required foundation for transistor theory, which is reviewed in the next chapter.

2-1 THE SEMICONDUCTOR DIODE

The semiconductor diode is a single-crystal structure in which both P-type and N-type semiconductor materials exist. Knowledge of the process of manufacturing such a device is not a prerequisite for understanding its operation; however, it is important to know that many of the devices and diagrams used to explain diode action are no more than convenient aids and should not be construed as physical facts.

One such aid is the frequently used PN junction schematic shown in Fig. 2-1. This figure shows that the P material, with its excess holes, and the N material, with its excess electrons, are each electrically neutral in their natural state. It is only when the two types of material are joined in a single crystal and the excess electrons from the N material combine with the holes of the P material that the potential barrier between the two materials is established. Theoretically, this barrier exists across a *depletion region* where no mobile carriers of either type exist as a result of the electron-hole pairing.

After the junction is formed, neither the P material nor the N material is electrically neutral since each has taken on (or released) charge from the other. It is easy to see from Fig. 2-1 that the formed junction has all of the characteristics of a charged capacitor, with the depletion region acting as the

The Semiconductor Diode

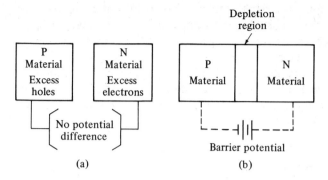

FIGURE 2-1. *P* and *N* material: (a) before junction is formed; (b) after junction is formed.

dielectric. It is the effective junction capacitance which becomes an important factor in all high-frequency operations of the diode.

The effects of connecting an external potential source to the PN junction are shown in Fig. 2-2. Forward biasing of the junction results in the free flow of current across the junction, with the external sources acting as an unlimited supply of carriers (electrons), thereby permitting a continuous recombination process. On the other hand, reverse bias is accomplished by connecting the external source with a polarity that reinforces the previously established potential barriers. With the increased potential barrier resulting from reverse bias, very little current flows across the junction, and the diode reacts like a very high resistance.

Note also in Fig. 2-2 the pictorial difference in the size of the depletion region under the two bias conditions. By controlling the potential barrier, one effectively controls the amount of junction capacitance; this is basically

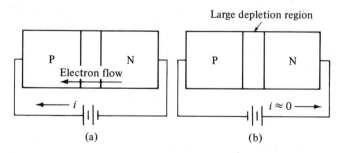

FIGURE 2-2. Biasing the *PN* junction: (a) forward bias; (b) reverse bias.

the principle of variable reactance device known as the varactor diode. For our purpose it is sufficient to note, at this time, that junction capacitance results because a depletion region exists and that there is a potential difference across this region.

In summary, the junction diode is a device that exhibits a very small (usually negligible) forward resistance and a rather large reverse or backward resistance. It has an internal capacitance that affects the high-frequency operation of the device.

2-2 THE DIODE EQUATION

It is possible to express mathematically the current that will flow across a PN junction. The equation is:

$$I = I_0(e^{k(V)/T} - 1), \tag{2-1}$$

where k is a constant for the type of diode used,
V is the applied voltage,
T is the absolute junction temperature,
I_0 is the reverse saturation current.

In Eq. (2-1), positive values of voltage represent forward bias, and negative values, reverse bias. Examination of the equation shows that, for a silicon diode (with $k = 5.8 \times 10^4$), the exponential term rapidly becomes much greater than unity; thus, the forward bias condition is one of a rapidly increasing exponential multiple of the reverse saturation current. On the other hand, the reverse bias condition with negative applied potential shows the constant term (1) to be dominant in the bracketed quantity; therefore, under reverse bias only the reverse saturation current I_0 flows. The reverse saturation current I_0 is also dependent on temperature. However, this temperature dependence has been omitted from Eq. (2-1) and will not be considered here. Typically, the reverse saturation current for a germanium diode is larger by several orders of magnitude than for a silicon diode. Values of less than .01 μA are not uncommon for reverse saturation current of a silicon diode at room temperature. Figure 2-3 is a graphical presentation of Eq. (2-1) and will be discussed in the next section.

2-3 THE CHARACTERISTIC CURVE FOR THE DIODE

A plot of the voltage-current relationship for a typical junction diode is shown in Fig. 2-3(a); Fig. 2-3(b) shows an idealized version of the same curve. The characteristic curve is divided into three parts by the breakdown voltage

The Characteristic Curve for the Diode

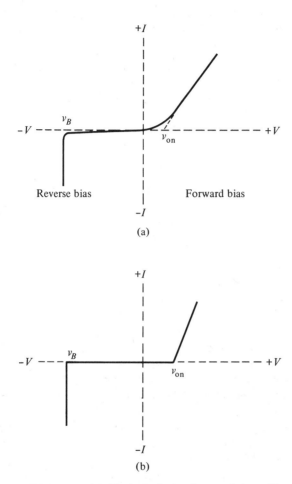

FIGURE 2-3. (a) Typical diode characteristics; (b) idealized version.

v_B and the turn-on or conduction voltage v_{on}. Although the general shape of the characteristic is similar for all diodes, there are variations from one specific diode to another. Generally, however, silicon diodes have a higher conduction potential than germanium diodes and a more linear reverse bias characteristic. A comparison between a silicon diode and a germanium diode is shown in Fig. 2-4.

Figures 2-3 and 2-4 both show that these diodes draw considerable current when reverse bias is increased beyond the breakdown voltage. In fact, if a normal diode is operated in this region, it will be permanently damaged;

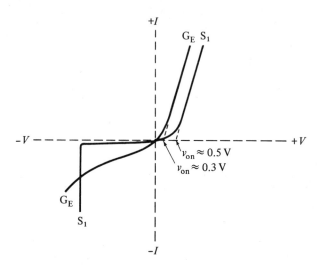

FIGURE 2-4. Comparison of germanium and silicon diodes.

therefore, the circuit designer must insure that the maximum inverse voltage to which a diode will be subjected is less than v_B. Diodes that can be cycled beyond v_B without damage are commonly called Zener or Avalanche diodes. They are most frequently used as voltage regulators because they adjust well to small voltage fluctuations that result from large current changes near the breakdown voltage. Equation (2-1) does not predict the diode operation in the breakdown region.

When the voltage across the diode is between v_B and v_{on}, the diode is considered to be reverse biased. Even though there is a small positive voltage between zero voltage and v_{on}, the current that flows is relatively small compared to that which would normally flow when the diode is fully forward biased. The actual determination of v_{on} is somewhat arbitrary; it is conveniently established by extending backward the linear part of the forward bias curve until it intersects the voltage axis. (See Fig. 2-3a.) Notice that in reverse bias the resistance of the diode is approximately

$$R_r = \left.\frac{\Delta v}{\Delta i}\right|_{v_B < v < v_{on}} \tag{2-2}$$

as indicated by the idealized Fig. 2-3(b), and is a large value (small slope). For a silicon diode, this value is typically on the order of several MΩ and, as we note from Fig. 2-4, it is considerably less for germanium diodes.

In the forward bias region the slope of the curve is rather large, indicating a low forward resistance. In this region,

$$R_f = \left.\frac{\Delta v}{\Delta i}\right|_{v_{on} < v},\qquad(2\text{-}3)$$

there is little difference between germanium and silicon diodes. Forward resistance of the diode is on the order of a hundred ohms, which in most applications is negligible compared to the other resistances in a given circuit.

2-4 DIODE RESPONSE TIME

When a diode in a circuit is reverse biased and the circuit input is adjusted to cause conduction, the response is not instantaneous. Both the magnitude of the reverse current and the speed of the input switching waveform are factors that determine how rapidly the diode will become forward biased and thus conduct fully. If the diode is slightly biased in the reverse direction, a considerable junction capacitance can be established. This capacitance will increase the response time in a switching situation. Normally, however, diode recovery from reverse bias to forward is negligible—typical times being less than 100 nsec.

Unfortunately, it is not so easy to neglect the events that occur when a conducting diode has to be switched off. Because conduction through the junction depends on the diffusion of minority carriers, a certain time elapses between the application of the switching potential and the time when reverse bias is actually accomplished. Part of this time is generally referred to as *storage delay time* and is that time required to reduce the number of minority carriers from the high level, which results from forward bias, to the equilibrium level, which exists if there is no bias. Finally, there is an additional time required to sweep the junction clean of minority carriers and establish the reverse bias potential. This time, usually referred to as the *transition time*, is completed when the junction capacitance has charged to the full value of the reverse voltage applied to the diode circuit (assuming negligible reverse current).

Figure 2-5 illustrates a simple circuit and the events that are described above. The time shown is typical; however, there are several variables that may affect it. Generally, it is best to establish the exact switching time experimentally.

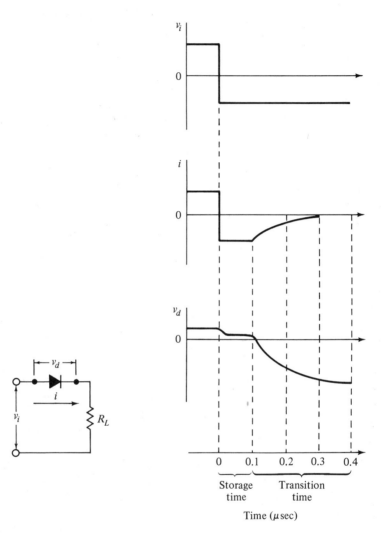

FIGURE 2-5. Response time in a diode circuit.

2-5 DIODE CLIPPING CIRCUITS

The diode clipper is one of the more basic, commonly encountered circuits. When employed in power supply circuits with sinusoidal signals it is referred to as a rectifying circuit. The half-wave rectifier or clipper, along with the pertinent circuit waveforms, is shown in Fig. 2-6.

Diode Clipping Circuits

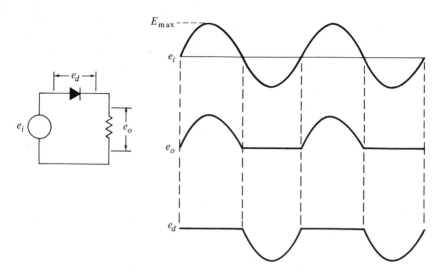

FIGURE 2-6. The half-wave rectifier.

A frequently used circuit variation is one where an additional potential source is used. As is obvious from Fig. 2-6, the output waveform (either the resistor voltage or the diode voltage) is dependent on whether or not the diode is conducting. Insertion of a battery as in Fig. 2-7(a) alters the output waveform by preventing diode conduction until the input voltage reaches the level of the battery.

Example 2-1 Sketch the output waveform of Fig. 2-7(a) if the circuit parameters are 1 kΩ, 10 V, and 15 sin ωt, for R, V, and e_i, respectively. Assume that D_1 is an ideal diode.

Solution: Diode D_1 is reverse biased until point B is more positive than point A. During this time the diode acts like an open switch, and e_o is 10 V. However, when e_i increases to a value that is slightly more positive than 10 V, the diode becomes forward biased, representing the closing of the switch; at this time e_o will follow the input voltage.

Since the input voltage will reach a peak value and then decline, the diode will be nonconducting until e_i equals 10 V. The output voltage waveform just described is shown in Fig. 2-7(a). The reversal of the diode or the battery is possible and will alter the output accordingly, as demonstrated in Example 2-2.

Example 2-2 Sketch the output waveform of the circuit of Fig. 2-7(b). The circuit parameters remain as in Example 2-1.

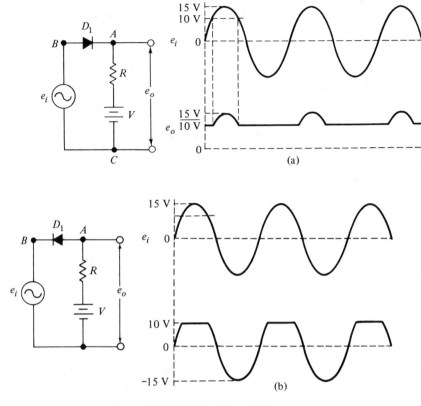

FIGURE 2-7. Biased clipper.

Solution: In this case the diode D_1 will be forward biased until point B is more positive than point A. During this biased "on" condition of D_1 the output voltage is e_i. At the time e_i reaches 10 V positive, D_1 opens and e_o is 10 V. This waveform is shown in Fig. 2-7(b). Observe that the positive peaks of the input signal are clipped and that the diode is conducting for most of the input signal period. The clipping level and conducting time of D_1 changes as the voltage V is varied.

The most frequent application of clippers in pulse circuitry is that in which trigger pulses of a particular polarity are not desired. Such a scheme is shown in Fig. 2-8. The RC differentiator produces both positive and negative spikes in response to the square wave input. If only one trigger per complete cycle of the input signal is desired, a clipper can be employed to eliminate either the positive or the negative pulses (as shown).

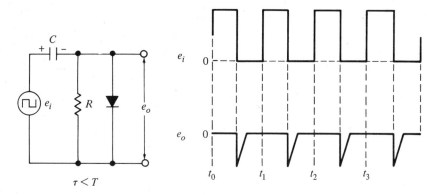

FIGURE 2-8. Differentiator with a diode clipper.

Summarizing, it is fair to say that diode clippers are among the simplest types of circuits used in pulse circuitry. It is important to remember, however, that the junction capacitance of the diode, which was ignored in all of the previous examples, can become a factor when high-frequency inputs or sharp spikes are encountered. Then, if the output is taken across the diode, there is a capacitive loading effect. If the diode is connected in series within the circuit and the output is taken across some other element, then the junction capacitance can couple high-frequency components across the diode; and this also will affect the output waveshape. So, for best results, it is desirable to drive the diode well beyond the conduction potential. The region of the nonlinearity of the diode characteristic causes rounding of the output waveform, as illustrated in Fig. 2-9, where diodes are employed to generate a square wave from a sinusoidal input. Throughout the discussion the ideal diode concept was applied, in that negligible voltage exists across the diode in a forward conduction mode; the resistance of the diode is also negligible in this mode of operation.

2-6 DIODE CLAMPS

Another basic application of diodes is in clamping circuits or, as they are sometimes called, dc restorers. These are circuits that always contain a capacitor and that have a large time constant relative to the period of the input signal. The back resistance of the diode itself can be the resistance in the RC time constant; however, more frequently there is a resistor in parallel with the diode as shown in Fig. 2-10.

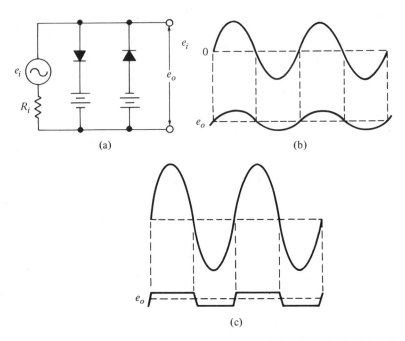

FIGURE 2-9. (a) A simple square wave generator; (b) with small signal input; (c) with large signal input.

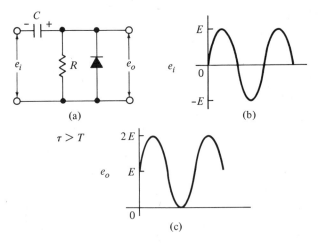

FIGURE 2-10. (a) The diode clamp; (b) input and output signals.

Diode Clamps

What occurs in a circuit such as Fig. 2-10 is that the "charge" time constant is very small, and the capacitor charges up to the peak value of the input signal; this charge is retained because of an extended "discharge" RC time constant. Since it would take a long time for the charge to leak off the capacitor, and since the input signal amplitude changes in a relatively rapid manner, then at any given time, according to Kirchhoff's voltage equation,

$$e_i - e_C - e_o = 0. \qquad (2\text{-}4)$$

But if e_C is charged to E with polarity as shown,

$$e_i + E - e_o = 0$$

and

$$e_o = e_i + E, \qquad (2\text{-}5)$$

which is the waveform shown in Fig. 2-10(c). Thus we see that the input sinusoid, which has no dc level, is clamped to a dc level and that no negative polarity exists. It is important to observe that the waveform was not distorted, as in the case of the clipper, but was just shifted to a dc reference voltage, hence a dc restorer circuit.

The action described above and shown in Fig. 2-10 is the final or steady state condition of a clamping circuit. The situation where the capacitor is initially uncharged must also be considered, for this is the case that exists when power is first applied to the circuit. In the first instant, current will flow freely if the input signal is negative-going. (See Fig. 2-11.) The output voltage will be zero because of the conducting diode. The capacitor will be accumulating charge in a manner consistent with the requirement that the sum of the voltages around the loop must be zero. During this first quarter cycle, the "charge" time constant is effectively small, because the conducting diode in parallel with R represents a small resistance, thus permitting the capacitor to charge rapidly with the input signal. After this quarter cycle the capacitor voltage remains fixed, as to move would require a leakage current in the direction opposite to that from which it charged. The diode cannot permit this and the resistance is too high to permit a significant loss of charge during the next half cycle. Thus, from this point on the capacitor voltage is acting like a battery voltage equal to $E_{\max}$, and the output is clamped. If, sometime later, there is a change in the input signal amplitude, then the circuit will respond by adjusting to a new dc level, while maintaining the negative peaks at zero volts. (See Fig. 2-12.) If the resistances were not included in the circuit of Fig. 2-10, then this readjustment of the dc level would, theoretically, not be possible. The time T required for the circuit to settle to the new level is determined by the RC time constants, as is shown by the exponential (dotted lines) in Fig. 2-12.

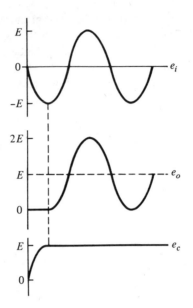

FIGURE 2-11. Initial waveforms in the diode clamp.

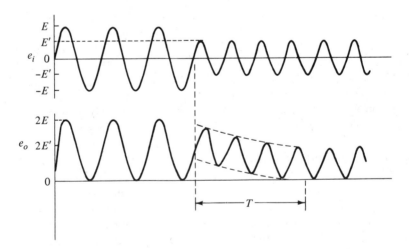

FIGURE 2-12. Effect of decreasing the input signal level to a clamping circuit.

2-7 CLAMPING THE NONSYMMETRICAL SIGNAL

The sinusoids discussed in the preceding section have a symmetry that is not always encountered. Consider the effect of the diode clamp on a nonsymmetrical square wave like that of Fig. 2-13(b). Qualitatively, we can make some observations about the output of a circuit like that shown in Fig. 2-10 when the nonsymmetrical signal is applied. When the positive part of the signal is impressed on the circuit, the diode will be reverse biased and the controlling circuit time constant will be RC. When the signal goes negative, the diode will conduct, and a much shorter time constant CR_f, where R_f is the diode forward resistance, will apply. The actual effect of the clamp on the signal will depend on the relation between these two time constants and the nonsymmetrical periods of the input signal. Just as for the previous (sinusoidal) case, there will be an initial situation, governed by the fact that

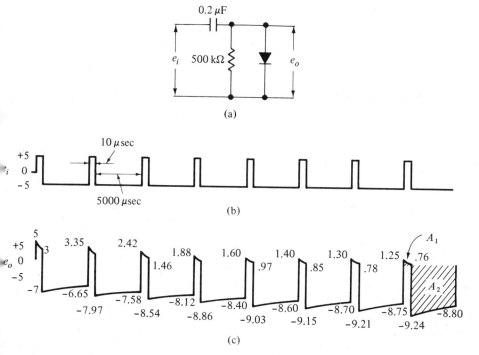

FIGURE 2-13. Circuit diagram and solution for the example problem.

the capacitor is originally uncharged, and a final or steady state condition, which is achieved shortly after the signal is applied. These points will be illustrated in the following example.

Example 2-3 Given the clamping circuit of Fig. 2-13(a) and the input signal in Fig. 2-13(b), calculate the output signal. Assume that, initially, the capacitor has no charge. Perform calculations through enough cycles to show what the steady state output will be. Assume that R_f is 100 Ω.

Solution: At t_0, 5 V are applied and impressed immediately across the parallel diode and resistor. The diode conducts because of the polarity of the signal. The time constant that governs the circuit operation is

$$\tau_1 = CR_f = (.2)(10^{-6})(100) = 20 \text{ } \mu\text{sec}.$$

During the first 10 μsec the output signal will decay as a result of the capacitor accumulating charge. After 10 μsec the output voltage will be

$$e_o = 5e^{-t_1/\tau_1} = 5e^{-10(10^{-6})/20(10^{-6})} = 5e^{-1/2} = 3.0 \text{ V}.$$

At this time, the input drops 10 V (from +5 to −5), and, because C cannot respond instantaneously, a 10 V drop in the output also occurs. (A Kirchhoff's loop equation with −5 V for e_i and −2 V on the capacitor is another method of determining the output voltage to be −7 V immediately after the switch at t_1.)

Now, with −5 V applied and 2 V on the capacitor the circuits will attempt to react. During this part of the cycle, however, the diode is reverse biased, and the parallel combination of the diode back resistance and the 500 kΩ is equivalent to the 500 kΩ by itself. The governing time constant is

$$\tau_2 = CR = (.2)(10^{-6})(500)(10^3) = 100,000 \text{ } \mu\text{sec}.$$

With this long time constant, there will be very little decay of the output voltage; however, as we shall see, it is not negligible in this problem. At t_2 the signal will have become

$$e_o = -7e^{-t/\tau_2} = -7e^{-5000(10^{-6})/100,000(10^{-6})} = -6.65 \text{ V}.$$

Next, the input jumps 10 V, which is again impressed on the output.

Successive calculations are similar to those above, with the starting voltage being appropriately adjusted each time the signal switches. The results of these calculations are shown in Fig. 2-13. Note that in Fig. 2-13(c) the output signal is clamped nearly, but not quite, at the zero level. This is a combined result of the choice of circuit components and the period of the input signal. As we shall see from the problems at the end of the chapter, varying degrees of clamping can be achieved. It can still be proven, however, that regardless of the input waveform, a relation between the part of the signal when the

Ground and Delay Clamping Circuit

diode is forward biased and the part when it is reverse biased does exist. This relation is shown in the last cycle of the waveform of Fig. 2-13(c) and is actually

$$\frac{A_1}{A_2} = \frac{R_f}{R}, \tag{2-6}$$

when A_1 and A_2 are the areas shaded in the illustration. This relation holds only when the steady state condition is reached in the circuit.

2-8 GROUND AND DELAY CLAMPING CIRCUIT

Many communications systems use AVC or AGC (automatic volume control or gain control) circuits to change the voltage gain of amplifiers as a function of input signal strength. This action is usually accomplished by varying the grid or base bias of the particular amplifier through the AVC or AGC common-line connection. In some instances it is desirable to delay this action until the input signal increases to a predetermined value, at which time the AVC circuit will operate. The response just described is that of a delayed AVC circuit. The ideas of ground clamping and delaying action are illustrated in the following example and in the circuit of Fig. 2-14.

Example 2-4 Determine the value of voltage V_1 that is required to switch the diode D_1 off in the circuit of Fig. 2-14. Assume that D_1 is an ideal diode.

Solution: The circuit can be redrawn as shown in Fig. 2-14(b) and

$$I_D = I_2 - I_1. \tag{2-7}$$

The current I_2 can be greater than or equal to I_1, but it will never be less than I_1. The switching of the diode occurs when the diode current I_D equals zero. Therefore,

$$0 = I_2 - I_1$$
$$I_1 = I_2$$
$$\frac{V_1}{R_1} = \frac{V_2}{R_2}$$

and

$$V_1 = \frac{V_2 R_1}{R_2}.$$

V_1 is often referred to as the switching voltage V_s, and as seen from the above equation, its value is determined by the circuit parameters.

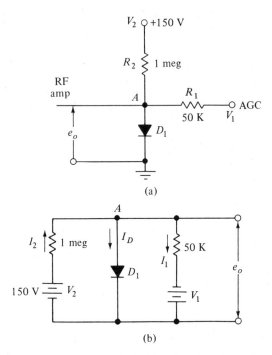

FIGURE 2-14. (a) Diode delay clamping circuit; (b) equivalent circuit to determine switching voltage.

$$V_s = \frac{V_2 R_1}{R_2} = -7.5 \text{ V}. \quad (2\text{-}8)$$

What this means is that V_1 must be at least -7.5 V for point A to go negative and turn the diode off. From a practical point of view, the output is essentially clamped at ground, and e_o will be delayed from changing until V_1 becomes equal to or less than the switching voltage V_s.

Review Questions

1. Explain the meaning of the *depletion region* as related to PN junction theory.
2. How many modes of operation can a PN junction diode be used in?
3. Explain in detail the characteristics of the diode for each mode of operation as stated in Question 2.
4. What is meant by forward and reverse bias?

5. What are the predominant current carriers in P-type semiconductor material?
6. Does operating a diode in the breakdown region necessarily damage the device? Explain.
7. What is the meaning of the factor I_0 in Eq. (2-1)?
8. What factors affect recovery time of a semiconductor diode?
9. Explain the meaning of the term *junction capacitance* as used in semiconductor theory.
10. In what material are the electrons a major current carrier of a PN junction? Explain.
11. What is meant by *storage time* as applied to a PN junction? Explain.
12. State in general terms the difference between diode clipping and diode clamping circuits.
13. Define the electric term *switching voltage* as it relates to pulse circuits.
14. Will decreasing the reverse voltage across a PN junction cause the junction capacitance to *increase* or *decrease*?
15. Explain the meaning of the "ideal diode" concept.

Problems

1. Determine the current I in the PN junction of Fig. 2-2(a) if the following values for I_0, V, and T/k are 15 µA, .1 V, and 26 mV, respectively.
2. Plot e_o of the circuit in Fig. 2-6 under two circuit conditions: (1) under the ideal diode concept and (2) if the diode has 100 Ω of forward resistance. $R = 500$ Ω, $e_{in} = 5 \sin \omega t$. Discuss the two waveforms.
3. Repeat Problem 2 for $R = 10$ kΩ.
4. Sketch V_o of Fig. 2-7 if e_{in} is a symmetrical square wave having a peak voltage of 30 V (± 15 V).
5. What would a dc voltmeter (VOM) indicate when connected across the output of the circuit in Problem 4?
6. Repeat Problem 4 by plotting the voltage across R (V_R) instead of e_o.
7. For what fraction of the input period is the diode in Example 2-1 conducting? The frequency of the input signal is 100 Hz.
8. What is the peak current in the diode of Problem 7?
9. Perform calculations through enough cycles to show what the steady state output will be for the clamping circuit of Fig. 2-13(a). The only circuit changes are 100 kΩ and ± 8 V for R and e_i, respectively.
10. Plot the output waveform of Fig. 2-10 if C, R, and R_f are .32 µF, 500 kΩ, and 100 Ω, respectively. e_i is a 100 V p-p or ± 50 V symmetrical square wave with a period of 16 msec.

11. What would a dc voltmeter (VOM) indicate if it were connected across the output of the circuit in Problem 10?
12. Refer to Fig. 2-14 and Example 2-4 to answer the following questions:
    (a) What is the current in the diode when V_1 is -4?
    (b) What is the current I_1 when V_1 is -4 V?
    (c) Repeat (a) and (b) for $V_1 = -7.5$ V.
    (d) Repeat (a) and (b) for $V_1 = -10$ V.
    (e) What is the new switching voltage V_s if R_1 is increased to 200 kΩ?
13. The diode of Fig. 2-14 is reversed. R_1 and R_2 are 20 kΩ and 100 kΩ, respectively. V_1 is an ideal ramp voltage changing from -100 V to $+150$ V with a slope of 1, and V_2 remains at 150 V. The period of the ramp is 1 msec. Plot e_o for two cycles of the ramp driving voltage.

Transistor and Vacuum Tubes

3

The most common active devices (that is, those capable of amplification) employed in switching circuits are transistors and vacuum tubes. In recent years there has been far greater application of solid state devices in the area of pulse and digital circuits, mainly because of the increased use of digital computers. Therefore, the treatment of the transistor is given preference in this chapter since it is singularly the most important device in the field. On the other hand, vacuum tubes have not been made obsolete, nor will they be in the near future, so we will devote time to their application too. Recent technological developments in the solid state field have provided other devices that are finding their way into common use, but the introductory level of this text requires that these devices be treated separately, even though they may be employed similarly to the transistor. Thus, a later chapter will be devoted to these devices.

3-1 THE JUNCTION TRANSISTOR

An NPN junction transistor amplifier using the common base configuration is shown in Fig. 3-1(a). Schematically, exactly the same circuit could appear as in Fig. 3-1(b), where the common connection (ground) between the negative side of V_{CC} and the positive side of V_{EE} is implied.

For normal amplifier operation, the emitter base junction is forward biased by the emitter supply V_{EE}, and the collector base junction is reverse biased by the collector supply V_{CC}. The selection of battery potentials and resistors establishes the quiescent, or steady state, operating condition. For Class A amplifiers this quiescent point would be established in the middle of the transistor's linear operating range, but in many pulse circuits the quiescent state will be either a heavy conduction (saturation) or no conduction (cutoff) state.

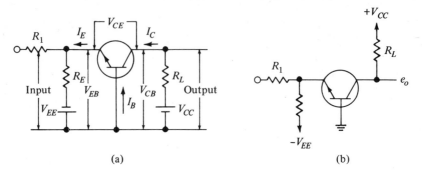

(a)   (b)

FIGURE 3-1. The common base configuration.

The application of a signal across the input terminals of the circuit of Fig. 3-1 will result in dynamic operation of the circuit. In the common base configuration, a positive-going input signal will result in a similar positive-going change in the output signal. Thus, a positive signal potential decreases the forward bias of the emitter base junction, resulting in fewer carriers for conduction across the collector base junction, less voltage drop across R_L, and, therefore, a rise in potential at the collector. When operated dynamically, a measure of the change in collector current for a change in emitter current is called the common base current gain, α (alpha)

$$\alpha = \frac{\Delta I_C}{\Delta I_E}\bigg|_{V_{CB}=k} \qquad (3\text{-}1)$$

In a junction transistor α is always less than unity.

Transistors are more frequently employed in the common emitter configuration, as shown in Fig. 3-2(a). Again, the more commonly observed sche-

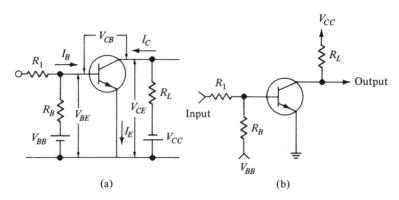

(a)   (b)

FIGURE 3-2. The common emitter configuration.

matic representation is shown in Fig. 3-2(b). The battery potentials are again arranged to provide forward bias of the emitter base junction and reverse bias of the collector base junction; however, in this configuration, the input signal is applied to the base lead. The current amplification factor β (beta), or the forward current gain in the common emitter configuration, is

$$\beta = \frac{\Delta I_C}{\Delta I_B}\bigg|_{V_{CE}=k} \quad (3\text{-}2)$$

and is much greater than unity.

As we can see, there is little difference between the common base and common emitter mode of operation; therefore, we can expect some relation between the current gains for the two modes. This relation may be developed as follows:

$$I_B = I_E - I_C \quad (3\text{-}3)$$

$$\beta = \frac{\Delta I_C}{\Delta I_B} = \frac{\Delta I_C}{\Delta I_E - \Delta I_C}$$

$$= \frac{\Delta I_C/\Delta I_E}{1 - \Delta I_C/\Delta I_E}$$

$$\beta = \frac{\alpha}{1 - \alpha}. \quad (3\text{-}4)$$

Similarly, we can prove that

$$\alpha = \frac{\beta}{1 + \beta}. \quad (3\text{-}5)$$

Example 3-1 Determine the β of a transistor that has a common base current gain α of .975.

Solution: From Eq. (3-4)

$$\beta = \frac{\alpha}{1 - \alpha} = \frac{.975}{1 - .975}$$
$$= 39.$$

If α was rounded off to .98 or .97, error would exist for β. Is this an objectionable error, or is it within limits that are tolerable? To answer this question one need only solve for β in each case.

Therefore,
$$\alpha = .98$$
$$\beta = \frac{.98}{.02} = 49$$

and
$$\alpha = .97$$
$$\beta = \frac{.97}{.03} = 32.$$

It is obvious from the above calculations that α should not be rounded off when solving for β by using Eq. (3-4).

For PNP transistors all of the foregoing comments apply; however, the battery polarities shown in Figs. 3-1 and 3-2 must be reversed to provide the correct bias.

3-2 CHARACTERISTIC CURVES FOR THE TRANSISTOR

A number of different characteristic curves for a transistor can be drawn, depending on the configuration used and the parameters of interest. Of most importance, however, are those that are referred to as the collector characteristics. A typical set is shown in Fig. 3-3.

The obvious reason for the choice of these particular sets of curves is their similarity to the plate characteristic curves for a pentode vacuum tube. The potential from collector to the common lead is analogous to the plate-to-cathode potential, while the collector current is analogous to the plate current in a tube. The input parameter is either the emitter current or base current, depending on the mode of operation used.

A point which should be emphasized when studying these curves is that the predominant factor that controls collector current in a transistor is base current, not collector voltage. If varying the voltage across a device does

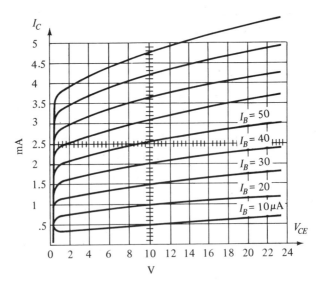

FIGURE 3-3. Collector characteristics for a typical transistor.

Leakage Currents

not affect the current in that device significantly, the device acts like a constant current source. This is the case of a transistor. The following example will help to illustrate this point.

Example 3-2 The characteristic curve of a given transistor is shown in Fig. 3-3. This transistor is used in an amplifier as illustrated in Fig. 3-2(a), and the operating point values for V_{CE}, I_B, and I_C are 10 V, 10 μA, and .5 mA, respectively. Which parameter change would result in the greater collector current change: (a) if the collector voltage were increased to 20 V and I_B remained constant or (b) if V_{CE} remained constant and I_B were increased to 20 μA?

Solution: The answer, as seen from the curve, is base current. Increasing base current from 10 μA to 20 μA doubled the collector current, whereas only a slight change occurred when the collector voltage was doubled.

For the PNP transistor the polarities are reversed. It would be most correct to draw the curves upside down and backward from those in Fig. 3-3; however, it is more common to simply label V_{CE} as $-V_{CE}$ and so on, and to draw the curves as shown.

3-3 LEAKAGE CURRENTS

In a common base circuit, when the collector base junction is reverse biased and the input (emitter) is open-circuited, some current will flow. This current is commonly referred to as I_{CO} or I_{CBO}, the reverse (leakage) current, and is highly sensitive to junction temperature.

Correspondingly, in a common emitter connection, when the input (base) lead is open-circuited and the collector base junction is reverse biased, a collector current will flow. Again, this is a leakage current and is referred to as I_{CEO}. To establish the relation between I_{CBO} and I_{CEO} for a particular transistor we note:

$$I_C \cong \alpha I_E + I_{CO}$$
$$= \alpha(I_C + I_B) + I_{CO}, \quad \quad (3\text{-}6)$$

then

$$I_C - \alpha I_C \cong \alpha I_B + I_{CO}$$

$$I_C \cong \frac{\alpha}{1-\alpha} I_B + \frac{1}{1-\alpha} I_{CO}$$

$$I_C \cong \beta I_B + (\beta + 1) I_{CO}. \quad \quad (3\text{-}7)$$

But, as previously stated, when the base is open-circuited ($I_B = 0$) the $I_C = I_{CEO}$ and

$$I_{CEO} \cong (\beta + 1)I_{CO}. \tag{3-8}$$

Thus the leakage current in the common emitter mode is many times $(\beta + 1)$ greater than in the common base mode.

3-4 THE HYBRID OR h PARAMETERS

To define the operation of a transistor circuit mathematically it is convenient to have a set of parameters that relate the information contained in the characteristic curves to an equivalent circuit. Several sets of parameters have been defined; however, the most convenient and frequently used set is that referred to as the hybrid or h parameters. These parameters are based on the assumption of linear operation of the transistor. They are sometimes called the small signal parameters because completely linear operation of the transistor can be achieved only with small signal operations and in the mid (linear) region of the characteristic curves. Although switching circuits frequently operate in a nonlinear fashion with very large signals, it is important to develop the small signal parameters in order to understand transistor operation in general. Furthermore, a switching circuit will generally operate in the linear region for some part of its switching cycle.

In developing the h parameters, the black box concept is used. The transistor amplifier is considered as a four-terminal device (see Fig. 3-4). Input and output voltages and currents are measured under short- and open-circuit conditions or are determined from the characteristic curves provided with the

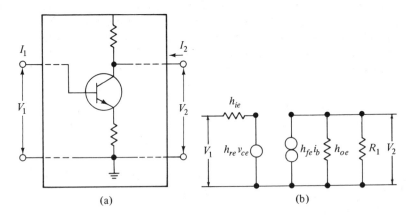

FIGURE 3-4. (a) Black box representation of common emitter amplifier; (b) equivalent circuit.

transistor. For the black box, the input and output voltages (v_1 and v_2), and the input and output currents (i_1 and i_2) are related by the hybrid parameters in the following manner:

$$v_1 = h_{11}i_1 + h_{12}v_2, \quad (3\text{-}9)$$
$$i_2 = h_{21}i_1 + h_{22}v_2. \quad (3\text{-}10)$$

From the above equations we note that if the output of the black box is short-circuited ($v_2 = 0$), then

$$h_{11} = \frac{v_1}{i_1}, \quad (3\text{-}11)$$

$$h_{21} = \frac{i_2}{i_1}. \quad (3\text{-}12)$$

Likewise, if we open-circuit the input ($i_1 = 0$),

$$h_{12} = \frac{v_1}{v_2}, \quad (3\text{-}13)$$

$$h_{22} = \frac{i_2}{v_2}. \quad (3\text{-}14)$$

By simply observing the units of volts, ohms, and amperes we see that

h_{11} is an impedance,
h_{21} is a current gain,
h_{12} is a voltage gain (transfer ratio),
h_{22} is an admittance.

For common usage the subscript notation is altered so as to be consistent with the type of amplifier and the specific parameter. Thus, for a common base amplifier we have

$h_{ib} = h_{11}$ is the input impedance parameter,
$h_{ob} = h_{22}$ is the output impedance parameter,
$h_{rb} = h_{12}$ is the reverse voltage transfer ratio,
$h_{fb} = h_{21}$ is the forward current transfer ratio,

and for the common emitter amplifier we have h_{ie}, h_{oe}, h_{re}, and h_{fe}.

Using the above information we can take Eqs. (3-9) and (3-10) and construct an equivalent circuit for the transistor amplifier as shown for the common emitter amplifier in Fig. 3-4. Equation (3-9) defines a loop that contains a voltage generator with an equivalent voltage of $h_{re}v_{ce}$ in series with an impedance of h_{ie}'. Equation (3-10) defines a second loop with an equivalent current generator of $h_{fe}i_b$ shunted by an admittance h_{oe}. With the use of this hybrid equivalent circuit, the standard equations for input impedance, output impedance, current gain, voltage gain, and power gain for a particular amplifier can be developed. Likewise, manipulation of the parameters may be accomplished to relate the common base hybrid parameters to the common emitter parameters. The student is referred to a standard transistor manual for the extensions of the material presented here.

3-5 THE TRANSISTOR AS A SWITCH

In the area of pulse and digital circuits the most frequent application of the transistor is with large signal operations, or, more specifically, as a switch. To better define the switching application, we will divide the characteristic curves for the transistor into three general regions and note that switching applications deal mostly with two of the regions—namely, cutoff and saturation. In these regions transistor operation is nonlinear; the small signal parameters and equivalent circuits developed in the preceding sections do not generally apply. For a common emitter configuration as shown in Fig. 3-5, the load line is divided into the three specific regions by points A and B. The active region, or region of linear operation, is where most amplifier circuits operate. In this region the state of the transistor is that the collector base junction is reverse biased and the emitter base junction is forward biased.

Cutoff in a transistor is achieved when both the emitter base junction and collector base junction are reverse biased. In this state, the leakage, or reverse

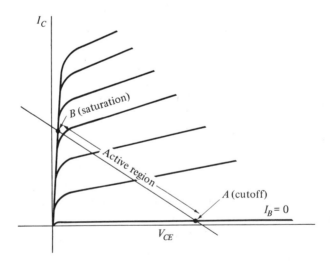

FIGURE 3-5. Transistor regions of operation.

The Transistor as a Switch

saturation current (similar to that in a reverse biased diode), is the only current that flows in the collector circuit. As we can see from Fig. 3-5, when the transistor is in cutoff, the collector emitter voltage V_{CE} is large and the collector current I_C is small. This condition is similar to an open switch so far as the load resistor R_L is concerned. Ideally, no current should flow through the resistor when the transistor is in cutoff; however, the transistor can only approximate the switch in this respect. It should be noted that cutoff is not achieved by opening the base lead, for with an open base it is the emitter current that flows across the collector base junction, not the reverse saturation current of that junction. Silicon transistors have a much lower leakage current than do germanium transistors. Furthermore, for $V_{BE} = 0$ a silicon transistor will be at cutoff, while to insure cutoff in a germanium transistor, a small reverse potential (on the order of .1 V) is required.

A transistor is in saturation when both the emitter base junction and the collector base junction are forward biased. The saturation region is shown in Fig. 3-5 as the region of operation to the left of point B on the load line. The total potential drop across both junctions is very small, and, corresponding to this, a high collector current nearly equal to V_{CC}/R_L will flow. A fact which is not evident from Fig. 3-5 but which is more clearly illustrated in Fig. 3-6 is that the saturation voltage $V_{CE(\text{sat})}$ has a range, depending upon the base current. Thus, where the curves appear as a single line at point B in Fig. 3-5, they are actually distinct when observed on the expanded scale of Fig. 3-6. This point is emphasized because a specification of the saturation resistance R_{CS}, where

$$R_{CS} = \left.\frac{V_{CE}}{I_C}\right|_{\text{saturation}}, \qquad (3\text{-}15)$$

can be somewhat misleading. Figure 3-6 shows that, in general, when the base current is greater than that represented by some minimum value ($I_{B(\min)}$), R_{CS} is nearly constant and approximately equal to the slope of the dotted line. For smaller values of base current R_{CS} is considerably larger, even though there is only a slight difference (on the order of a millivolt) in $V_{CE(\text{sat})}$. Generally, values for R_{CS} on the order of tens of ohms are not uncommon.

When the transistor is in saturation and operated in a circuit similar to that of Fig. 3-6(c), it is equivalent to a closed switch. The current through the load is $V_{CC}/R_L + R_{CS}$—nearly identical to V_{CC}/R_L, which would be expected for an ideal switch.

Example 3-3 Determine the values for I_B and I_C of Fig. 3-6(a) if R_1, R_B, R_L, V_{CC}, and β are 10 kΩ, 30 kΩ, 5 kΩ, 20 V, and 50, respectively. V_{in} is a square wave pulse of $+10$ V, and the transistor is a silicon device.

44   **Transistor and Vacuum Tubes**

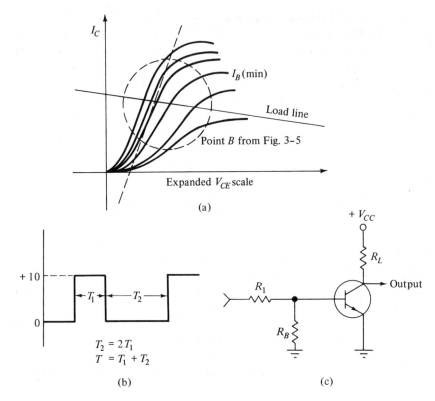

FIGURE 3-6. Transistor saturation region.

Solution: Two circuit conditions must be considered when solving the problem. First, when the input voltage is 0 V the transistor should be recognized as being turned off, and I_B is zero; therefore, I_C is zero (assuming negligible leakage current). The input voltage pulse then rises to $+10$ V and turns the transistor on, the $I_{B(on)}$ circuit is

$$I_{B(on)} = \frac{V_{in} - V_{BE}}{R_1} = \frac{+10 - .6\text{ V}}{10\text{ k}\Omega}$$

$$= \frac{9.4\text{ V}}{10\text{ k}\Omega} = .94\text{ mA}$$

and

$$I_C = \beta I_B.$$

Response of the Transistor Switch

However, a note of caution is in order at this point. The above equation is only true when the circuit operation is in the small signal or linear mode. That is, β is the small signal current gain parameter. Observing from the characteristic waves of Fig. 3-6, it should be noted that when a transistor is driven into saturation, the β of the device decreases. The device is referred to as operating in the "forced β" mode.

Observe that if
$$I_C = \beta I_B = (50)(.94 \text{ mA})$$
$$= 47.00 \text{ mA}$$
and
$$V_{CC} = I_C R_L + V_{CE}$$
$$20 = (47 \text{ mA})(5 \text{ k}\Omega) + V_{CE}$$
$$20 \neq 235 + V_{CE}.$$

This condition cannot exist, and the conclusion is that the transistor is driven into saturation. Therefore,

$$I_C = \frac{V_{CC}}{R_L + R_{CS}} \approx \frac{V_{CC}}{R_L} = \frac{20}{5 \text{ k}\Omega} = 4 \text{ mA}$$

and β (forced)

$$\frac{4 \text{ mA}}{.94 \text{ mA}} \approx \frac{4 \text{ mA}}{1 \text{ mA}} = 4.$$

3-6 RESPONSE OF THE TRANSISTOR SWITCH

It should be obvious that although the transistor may be switched between saturation and cutoff in the common base configuration, the common emitter is more desirable because a small input (i_b) controls a large output (i_c). A logical question is, "How fast will the output switch when the input is driven in a manner to cause switching?" The answer to this question depends on many factors, among which are the input signal itself, junction capacitance, minority carriers, storage delay, and the external circuitry. Consider Fig. 3-7, which shows a typical response for a transistor in the common emitter configuration when the input signal (assumed ideal) causes it to switch between cutoff and saturation. Prior to t_0 the transistor was maintained in cutoff by the negative potential (assuming an NPN transistor) applied to the base. We note from the diagram that the transistor does not respond instantaneously to the switch from $-V$ to $+V$ at the input. The time interval, t_0 to t_1, commonly called the turn-on delay, is due both to the transistor junction capacitances, which have changed to cutoff potential, and the definite time required for carriers to traverse the base region by the diffusion process. Thus we note

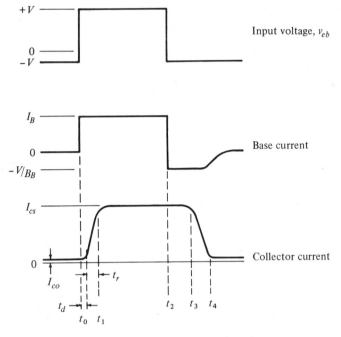

FIGURE 3-7. Illustration of transistor response.

that the physical construction of the transistor is a major factor in the determination of its response.

The turn-on delay time is usually considered as two separate parts, the delay time and the rise time. The delay time t_d is that time between the instantaneous change in base current and the time when the transistor enters the active region. In other words, during the delay time, the emitter base junction capacitance is charging toward the forward bias condition from its previous reverse bias. Thus, we see that the delay time can be made very small by combining the proper choice of transistor (construction) with a design that insures that the stage is not driven excessively into cutoff.

The rise time t_r begins when the transistor enters the active region and is generally considered complete when the collector current has increased to 90 percent of its final value. It is dependent upon both transistor construction and circuit design, as the commonly accepted expression for t_r illustrates:

$$t_r = h_{FE} \left(\frac{1}{\omega t} + kR_L C_C \right) \ln \frac{h_{FE} I_B}{h_{FE} I_B - 0.9 I_C}, \qquad (3\text{-}16)$$

where ωt = current gain-bandwidth frequency, that frequency where the magnitude of $\beta = 1$,
h_{FE} = large signal (dc) β,
I_B = base driving current,
C_C = output capacitances (collector junction plus external circuit),
R_L = collector circuit load resistance,
k = factor dependent upon transistor construction; generally, $1 \leq k \leq 2$.

A thin base region and low emitter junction capacitance both affect ωt and are very important factors in the design of switching transistors. Small load resistors provide for improved rise times at the expense of voltage swing at the collector. Large values of base drive current decrease the rise time, but as we shall see shortly, the extent to which the transistor is driven into saturation affects the turn-off time. Because of this, it is common to see an input resistor shunted by a capacitor (Fig. 3-8) in the base lead of a transistor switch. During the rise time period (when the input is initially applied), the capacitor bypasses the resistor and the base current is high. When the capacitor charges to the input voltage, the base current is reduced, as is the extent to which the transistor is driven into saturation.

At the trailing edge of the driving signal we note (from Fig. 3-7) that there is also a turn-off delay time. The turn-off delay is usually considered to consist of the storage time and the fall time. The storage time t_s is the same situation described previously for the junction diode. (See Fig. 2-5.) That is, because of saturation, the excess minority carriers in the base and collector regions must be swept out before current across the junction is reduced.

Once the transistor comes out of saturation, the analysis for the fall time t_f parallels that for the rise time. When the transistor is in the active region,

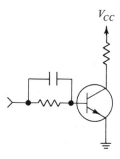

FIGURE 3-8. An input circuit improves rise time.

we can expect logically the same factors to affect its response whether the drive current is increasing or decreasing.

Considering the turn-off delay, an important factor is apparent. The storage delay time can be reduced to zero simply by avoiding saturation. In circuits where the trailing edge response is important, this is a common technique. The most common of such circuits are the nonsaturating switch and the nonsaturating flip flops, to be discussed later. The price one pays for the improved response time is an increase in power consumption, for we need only investigate the curves of Fig. 3-3 to note that more power is dissipated in the transistor when it is in the active region than when V_{CE} is small, as it is the saturation region.

3-7 VACUUM TUBES

Many years before the appearance of the transistor the vacuum tube was considered the basic electronic device, and, as a matter of fact, all of the circuits considered in this text were conceived and operational before transistors became commercially available. The natural advantages of the transistor, however, are responsible for the implementation of pulse and digital circuits on a large scale, and thus it was the transistor that led to the widespread development and use of computers. It is still important, however, to consider the vacuum tube as a practical device—as long as we recognize its limitations and advantages. Briefly then, we will consider the triode and pentode here, and throughout the text we will show vacuum-tube applications that parallel the specific transistor circuits. The single exception to this will be circuitry of the vacuum diode, for almost universally the solid state diode has replaced the vacuum-tube counterpart.

3-8 THE TRIODE AS A SWITCH

Like the transistor, the triode has an active region, a saturation region, and a cutoff region. A typical set of characteristics is shown in Fig. 3-9. When a triode is conducting, it can be driven beyond plate current cutoff simply by applying a negative voltage to the control grid, provided the voltage is of sufficient magnitude. As long as the grid is negative with respect to the cathode, a negligible amount of grid current will flow; then there is virtually no power drain from the driving or controlling source. Thus, in either the active or the cutoff region there is excellent isolation or decoupling between the inputs and outputs of the triode circuits. If the input signal does drive the grid positive with respect to the cathode, then grid current will flow, and

The Triode as a Switch

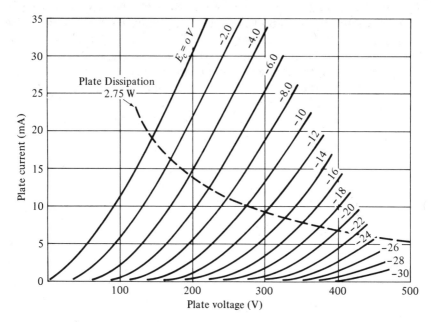

FIGURE 3-9. Plate characteristics for the 12AU7 triode.

the grid and the cathode will form a conducting diode. The amount of grid current will be determined by the amplitude of the driving signal applied to the grid and the source impedance of the driver.

Figure 3-10 shows a common connection for the grid leak in pulse circuits. It is referred to as "clamping the grid" because the grid voltage is maintained at nearly the same potential as the cathode. If the value of R_g is large, on the

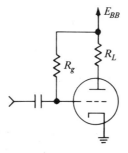

FIGURE 3-10. Grid clamping circuit.

order of 1 MΩ, and a small amount of grid current is flowing, then the grid potential will be very small because R_g forms a voltage divider with the effective resistance of the grid cathode path. This effective resistance is small compared with R_g. Thus, the potential from ground to the grid is very small, or, in other words, the grid and cathode are at nearly the same potential. Clamping in the grid cathode circuit in this manner is identical to clamping as discussed in the preceding chapter.

It is not necessary to return the grid to a high potential to clamp the grid,

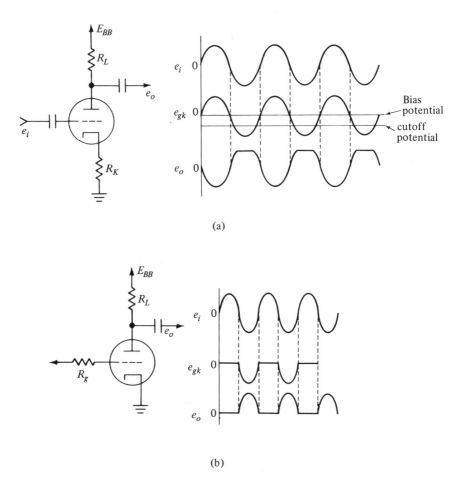

FIGURE 3-11. (a) The triode clipper, and (b) the grid current limiter.

since whenever a significant amount of grid current flows, the grid cathode potential is small. Grid current limiters make use of this feature by clipping the input signal when it exceeds zero volts. This is illustrated and emphasized in Fig. 3-11(a) because it is different from triode clipping, which is illustrated in Fig. 3-11(b). In the triode clipper, the tube is biased so that the negative portion of the input signal drives the tube into cutoff and there is no appreciable grid current or clamping during any part of the cycle. In the grid current limiter there is grid current for a part of each input cycle.

Finally, a triode can be driven into saturation when so much plate current flows that the plate voltage can go no lower. This is commonly called *bottoming* and is achieved by driving the grid with a positive signal but employing a low impedance source so that clamping does not take effect. The large grid leak resistor is also eliminated.

3-9 PENTODES

An idealized set of curves for a pentode is shown in Fig. 3-12. The physical construction of the tube results in a high plate resistance, which is reflected in the nearly horizontal characteristics.

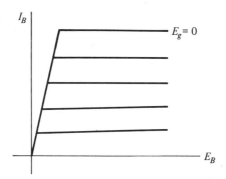

FIGURE 3-12. Idealized pentode characteristics.

The switching applications of the pentode are similar to those of the triode except in those circuits where the extra grids in the pentode are used separately for gating and control. Miller integrators and Phantastrons are circuits of this type that we will discuss in Chapter 7.

Review Questions

1. State the proper bias conditions for the junctions of a transistor when it is operating as a linear amplifier.
2. Why are *h* parameters sometimes called small signal parameters?
3. State the proper bias conditions for the junctions of a transistor when it is operated in the saturation mode.
4. Define the term *hybrid parameter*.
5. State the common collector hybrid parameters for the general parameters h_{11}, h_{12}, h_{21}, and h_{22}.
6. Define the term *collector saturation resistance*.
7. State the proper bias conditions of each function of a transistor if cutoff is to be achieved.
8. Define the electronic terms I_{CEO} and I_{CBO}.
9. State two basic factors that affect the high-frequency response of a transistor.
10. Define the term *rise time*.
11. What are the major factors that affect the rise time of a transistor switch?
12. What is the significance of having a thin base region in a switching transistor?
13. Increasing the load resistance of a particular transistor switch would affect what portion of the output waveform?
14. Why is more power dissipated in a transistor when it is operated in the active region of the device than when it is operated in saturation?
15. Why do the collector characteristic curves of a transistor resemble the plate characteristics of a pentode rather than those of a triode?
16. Why are transistors preferable to vacuum tubes in digital computers?
17. Explain how to solve for β by using a collector characteristic curve.
18. Why is I_{CBO} much less than I_{CEO}?
19. If the base current is increased, what effect does this have on the output impedance of a transistor? Why?

Problems

1. Determine the values for α of each transistor if their β's are 24, 49, 60, 99, and 200.
2. Using the alpha values obtained in Problem 1 as the independent variable, plot the β and α points to produce a curve.

Problems

3. What significant statements can be made from analyzing the β vs α plot obtained in Problem 2?
4. Determine I_B and I_C of the circuit in Fig. 3-2 if the circuit parameters are: $V_{BB} = 10\text{ V}$, $R_B = 100\text{ k}\Omega$, $R_L = 2\text{ k}\Omega$, $V_{CC} = 16\text{ V}$, $\alpha = .98$, and $V_{BE} = .3\text{ V}$. R_1 is grounded.
5. If the resistor R_B in Problem 3-4 were reduced and all other values remained constant, would the rise time of the circuit increase, decrease, or remain constant? Discuss your answer.
6. What would be the effect on rise time of a given transistor switch if the load resistance increased? Discuss your answer.
7. What are the V_{CE} and V_{CB} voltages of the circuit in Problem 4?
8. (a) Plot the output voltage waveform of the problem in Example 3-3.
   (b) What device parameter must be specified quite accurately when a transistor is turned off?
9. What would a dc milliammeter indicate if it were placed in series with the 5 kΩ load resistor in the circuit of Example 3-3.
10. Determine the small signal parameters h_{fe} and h_{oe} from the graph of Fig. 3-3 for constant values of $V_{CE} = 10\text{ V}$ and $I_B = 20\text{ }\mu\text{A}$.
11. Given the circuit of Fig. 3-2 and the input waveform of Fig. 3-7, the input waveform is a symmetrical waveform with $-V = 0$ and $+V = 12$. Assume the circuit to be a transistor switch; the parameters are: $R_1 = 10\text{ k}\Omega$, $R_B = 20\text{ k}\Omega$, $V_{BB} = -3\text{ V}$, $R_L = 4\text{ k}\Omega$, $V_{CC} = 20\text{ V}$, $V_{EB(\text{on})} = .6\text{ V}$, and $I_{CBO} = 15\text{ }\mu\text{A}$.
    Find: (a) I_B
    (b) $V_{BE(\text{off})}$
    (c) I_C
    (d) Plot the output waveform.
12. Place an 8 kΩ resistor from collector to ground of the circuit in Problem 11 and plot the output waveform.

Logic Gates and Transmission Gates

4

By definition, a gate circuit is one that yields an output only under specified conditions. When the specified conditions constitute a logic function, as is commonly the case in digital computers, we designate the circuit a *logic gate*. If the specified condition(s) is simply a control signal, which changes the state of a circuit to permit the transmission of a signal that would otherwise be blocked, we call the circuit a *transmission gate*. The two circuits are not always very different, and sometimes the only distinction between a logic gate and a transmission gate exists in the intended use for the circuit. Gate circuits are commonly used in all fields of electronics (they are by no means limited to computers) and make use of a wide variety of components to achieve the end result. We will be concerned in this chapter with the most common types of gate circuitry that employ diodes, transistors, and vacuum tubes.

4-1 LOGIC

Although it is not the intent to study logic at this time, its terminology and basic principles have become firmly entrenched in the associated electronic circuitry. Therefore, a brief discussion of logical principles will precede the discussion of several circuits.

The combination of logic with electronics results in a two-valued system. That is, we will usually be concerned with truth or the negative of truth: on or off, up or down, 0 or 1. The electronics components will require that the logic statement be made in terms of this two-valued, or binary, system. Three basic logic propositions can be used to establish truth. These propositions are AND, OR, and NOT.

The AND proposition is true, if, and only if, all components of the proposition are true. Thus, the statement *It is raining and it is cold* is true only if both portions of the postulate (rain, cold) are true. Otherwise, the statement is false. An electronic circuit that corresponds to the AND proposition is shown in Fig. 4-1. Suppose that when it is raining, a positive voltage is applied to A

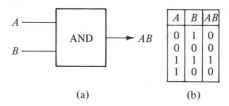

FIGURE 4-1. (a) AND gate; (b) truth table.

and when it is cold, a positive voltage is applied to B. The electronics components in the box will be so arranged that the output will be positive if, and only if, A and B are both positive. A truth table can be constructed for the AND circuit; it appears in Fig. 4-1.

The ones represent true statements and the zeros represent false. Although each input is limited to a bivalued function, the number of inputs is not limited to two. However, where the logical statement can have an unlimited number of postulates, the electronics is more limited, as should be expected.

The circuit for the OR function, along with the applicable truth table, appears in Fig. 4-2. The statement is true if either the A or the B portion of

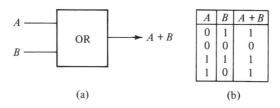

FIGURE 4-2. (a) OR gate; (b) truth table.

the postulate is true. (Form your own postulate.) Again, the truth table tells the story if we assume ones to represent truth and zeros the negative of truth. Like the AND circuit, the OR circuit is not restricted to two inputs.

The NOT function is simply the negation of a statement. Thus, if we let a one represent truth and a zero not truth, the circuit provides outputs as depicted in Fig. 4-3. Symbology normally associated with the logic so far discussed is simple:

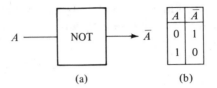

FIGURE 4-3. (a) The inverter or NOT circuit; (b) truth table.

$$\text{AND} = AB \quad = (A \text{ and } B)$$
$$\text{OR} = A + B = (A \text{ or } B)$$
$$\text{NOT} = \bar{A} \quad = (\text{not } A).$$

Example 4-1 Note the circuit of Fig. 4-4 and the associated truth table. Analyze the circuit to verify the truth table.

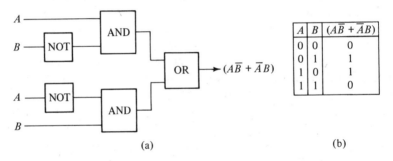

FIGURE 4-4. (a) The exclusive OR circuit; (b) truth table.

Solution: The circuit is commonly called the "exlusive OR." That is, an output exists for A and NOT B or NOT A and B, but no output exists for the A and B input as was the case for the simple "inclusive OR" shown in Fig. 4-2. The symbolic equation for the "exclusive OR" circuit is

$$\text{TRUTH} = (A\bar{B} + \bar{A}B).$$

The combination of the NOT circuit and either the AND or the OR circuit provides two gates called the NAND gate and the NOR gate. Many computers employ NAND-NOR logic, which is basically diode-transistor logic, (DTL), as opposed to the simple diode logic (DL) usually employed in the AND-OR circuits.

4-2 THE AND GATE

A typical AND gate employing diodes is shown in Fig. 4-5. The circuit makes use of positive logic, that is, an input (one) is represented by a positive voltage, and a no input (zero) is represented by a zero voltage or ground. In the circuit shown, we assume the combined forward resistance of the diode;

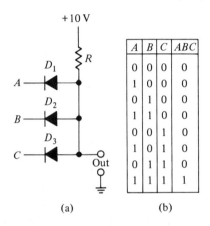

FIGURE 4-5. (a) Diode AND gate; (b) truth table.

each input source resistance is negligible compared to R. We further assume that the reverse biased diodes draw negligible current. Then, when any of the inputs is at the ground level (0 V), the corresponding diode will conduct, and the output will be the small (near zero) voltage developed across the conducting diode. Even if each of the other inputs is at the $+10$ V level, the shunting effect of the conducting diode determines the output voltage. Should all of the inputs be at the 10 V level, however, then all of the diodes would be reverse biased and the output voltage would be $+10$ V. Thus, the truth table depicts the circuit operation, with 1 representing the 10 V level and zero representing ground.

Example 4-2 What is the current in diode D_2 of Fig. 4-5 if all inputs are at ground (0)? Assume that $R = 5\text{k}\Omega$, $D_A = D_B = D_C$, and all are silicon devices.

Solution: With all inputs at ground,

$$I_T = \frac{V_{CC} - V_D}{R} = \frac{10 - .6 \text{ V}}{5 \text{ k}\Omega}$$

$$= \frac{9.4}{5 \text{ k}\Omega} = 1.88 \text{ mA} \approx 2 \text{ mA}$$

and

$$I_T = I_A + I_B + I_C.$$

Therefore,

$$I_B = \frac{I_T}{3} = \frac{1.88 \text{ mA}}{3} = .626 \text{ mA}.$$

4-3 THE OR GATE

Like the AND gate, the OR gate consists of a series of diodes that accept inputs and that, for certain combinations of inputs, provide an output. Figure 4-6 shows a typical OR gate (again, assuming positive logic) along with the truth table that applies.

The operation of the circuit is quite simple. If either input, A or B, is one, then the output is one. In this instance, a one could correspond to a positive voltage of some fixed value. Note that if a one is applied to both A and B, only one of the diodes will conduct if one of the inputs is slightly different from the other. The greater of the two inputs will appear at the output and the same voltage will hold the other diode in cutoff. In any event, the output will be less than the input by an amount equal to the voltage drop across the conducting diode.

In both the AND gate and the OR gate depicted above, the effects of diode junction capacitance, capacitance at the output of the circuit (wiring capacitance plus input capacitance to the next stage), and diode response and

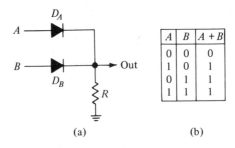

FIGURE 4-6. (a) Diode OR gate; (b) truth table.

The NOT Circuit

recovery times have been neglected. All of these factors become more important as the operational speed requirement is increased.

Example 4-3 The circuit of Fig 4-6 has inputs of $+10$ V and $+5$ V for A and B, respectively. R is equal to 10 kΩ and $D_A = D_B$, which are germanium devices. Determine the current in R and the voltage across D_B for these inputs.

Solution: Note that D_A will conduct and

$$I_R = \frac{V_{in} - V_D}{R} = \frac{10 - .3 \text{ V}}{10 \text{ k}\Omega} = .97 \text{ mA}.$$

If

$$V_{out} = +9.7 \text{ V},$$

then

$$V_{\text{across diode } B} = V_B - V_{out} = +5 \text{ V} - 9.7 \text{ V} = -4.7 \text{ V}.$$

The conclusion is that the anode of D_B is negative and therefore reverse biased; hence, no current flows.

4-4 THE NOT CIRCUIT

The NOT circuit is commonly called an inverter because its sole function is to provide an output that is the inversion of each input. Because an active device (transistor) is employed, the designer will use the inverter to amplify and reshape the pulse, which may have deteriorated after having passed through diode gates. The NOT circuit is a single-stage transistor circuit biased at cutoff. The application of a signal to the input causes the stage to saturate. In Fig. 4-7, the inverter output is V_{CC} volts when a zero input is applied. Since zero represents ground, V_{BB} holds the stage at cutoff and the output is V_{CC}. V_{CC} represents the one level in the circuit. The application of a one to the input of the inverter brings the stage from cutoff to saturation; the output becomes 0 V, representing zero. Note that if the input one had to be somewhat deteriorated in amplitude, the output could still reflect the proper levels for zero and one if the stage is designed to permit saturation with something less than the normal one level at the input.

The serious student of this technology should be familiar with the circuit theory of gates as well as their logical function. A detailed analysis as illustrated by the following example will provide greater understanding of this circuit and of future circuits in the book.

Example 4-4 Consider the circuit of Fig. 4-7 and assume that the transistor is an ideal device with no I_{CBO} leakage current. The circuit parameters

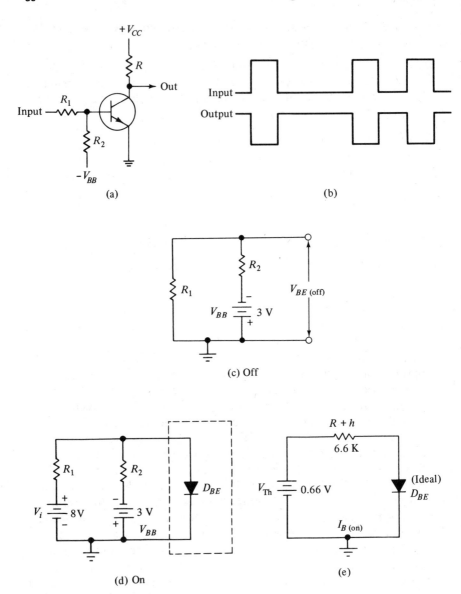

FIGURE 4-7. (a) Transistor NOT circuit; (b) input-output waveforms; (c) OFF equivalent circuit; (d) ON equivalent circuit; (e) Thévenins equivalent circuit.

The NOT Circuit

are: $R = 4\ \text{k}\Omega$, $R_1 = 20\ \text{k}\Omega$, $R_2 = 10\ \text{k}\Omega$, $V_{CC} = 24\ \text{V}$, $V_{BB} = -3\ \text{V}$, and V_{in} is swinging from 0 to $+8$ V, which is sufficient to saturate the device.

Find: (a) $I_{B(on)}$
      (b) $V_{BE(off)}$
      (c) $I_{C(sat)}$

Solution: It should be understood that there are two circuit states that need investigating. These two are the on and off states of the transistor. When V_{in} is 0 V the $-V_{BB}$ supply biases the transistor off, and the off equivalent circuit is illustrated in Fig. 4-7(c). The voltage appearing across R_1 is $V_{BE(off)}$. Therefore,

$$V_{BE(off)} = IR_1 = \left(\frac{-V_{BB}}{R_1 + R_2}\right) R_1$$

$$= \frac{-3}{30\ \text{k}\Omega}\, 20\ \text{k}\Omega = -2\ \text{V}.$$

The -2 V is sufficient to reverse bias the emitter base junction. Therefore, with no collector current present, the collector voltage is high ($+24$). The transistor is switched on when V_{in} is $+8$ V; the on equivalent circuit is shown in Fig. 4-7(d). This circuit is simplified by using Thévenin's theorem, as seen in Fig. 4-7(e). Therefore,

$$V_{th} = \frac{(V_{in} + V_{BB})}{(R_1 + R_2)} R_2 - V_{BB}$$

$$= \left(\frac{11}{30\ \text{k}\Omega}\right) 10\ \text{k}\Omega - 3$$

$$= +.66\ \text{V}$$

$$R_{th} = \frac{R_1 R_2}{R_1 + R_2} = 6.6\ \text{k}\Omega$$

and

$$I_{B(on)} = \frac{V+h}{R+h} = \frac{.66\ \text{V}}{6.6\ \text{k}\Omega} = 100\ \mu\text{A},$$

observing from the analysis that the base emitter diode is just forward biased. Any reduction in the input pulse would not turn the transistor on, especially if the diode junction voltage is considered. The β of the device is large enough for the 100 μA base circuit to drive the transistor into saturation. Therefore,

$$I_{C(sat)} = \frac{V_{CC}}{R} = \frac{24\ \text{V}}{4\ \text{k}\Omega} = 6\ \text{mA}.$$

The output voltage is close to ground, or 0 V. Notice that the output waveform is an inversion of the input. That is, when V_{in} is low or ground the output V_{CC} is high, and the circuit responds in the opposite way when the input reverses.

4-5 OTHER FORMS OF LOGIC CIRCUITS

Other types of logic circuits, along with various combinations of the ones discussed here, are in use today. Negative logic is identical to positive logic except that all polarities are reversed. Diode directions must be reversed and PNP transistors substituted for NPN. Resistor-transistor logic (RTL) makes use of resistive input circuits to transistors instead of diodes. Direct current transistor logic (DCTL) employs transistors only in a direct coupled fashion, where the only other circuit components employed are those required to maintain bias. Integrated circuit technology has added to the logic family a multiemitter transistor gate producing transistor-transistor logic (TTL). The interested student is referred to the great number of texts devoted solely to digital systems and logic.

4-6 TRANSMISSION GATES

Transmission gates differ from logic gates in that their purpose is to provide an output when a control signal is applied. Generally, it is required that the output be a reasonable reproduction of the input; for this reason, transmission gates are sometimes referred to as linear gates. We shall see that in some instances the actual circuitry of the transmission gate is identical to that of a logic circuit, while in other cases logic gates like we have just studied cannot satisfy the requirements for transmission gates. Thus, it is not always the actual circuitry that is responsible for the distinction between gates, but rather the intended use for a given circuit.

The simplest transmission gate conceivable employs a mechanical switch, as illustrated in Fig. 4-8. The control signal consists of the actuation of the

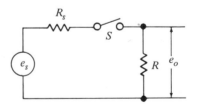

FIGURE 4-8. Simple transmission gate.

switch at the required time. Of course the circuit is not very practical except when the switch is of the electro-mechanical type (relay), and even these have severe limitations compared to the electronic switches that we intend to study here. (The advantages and disadvantages of mechanical switches have been discussed in Chapter 3.)

4-7 THE UNIDIRECTIONAL DIODE GATE

Corresponding to the circuit of Fig. 4-8, we have the unidirectional diode gate of Fig. 4-9. In this circuit, the control signal would consist of a positive voltage sufficient to override the reverse bias on the diode. Under such a condition, the diode would be conducting and would represent a low-resistance path to a positive-going signal voltage. In the absence of a control voltage, the applied signal would be greatly attenuated, assuming, of course, that the signal voltage is not large enough to override the reverse bias by itself.

The fact that both the signal voltage and the control voltage combine makes this circuit useful as a threshold gate. By adjusting the amplitude of the control voltage, the designer can impose a minimum (threshold) amplitude requirement for the signal voltage. The coincidence of a signal with the control signal will place the signal on the pedestal formed by the control voltage. This is a feature commonly used to eliminate base-line noise in certain applications; it is illustrated in Fig. 4-10(a). Note the similarity between this particular transmission gate and the logic AND gate. An output is obtained only when two coincident inputs exist. The emphasis with the transmission gate is on reproduction of the signal voltage, whereas with the logic AND gate, the emphasis is on the production of a proper signal level. Should the transmission of a signal depend on more than one coincident function,

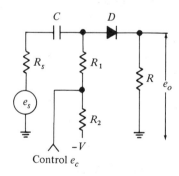

FIGURE 4-9. Unidirectional diode gate.

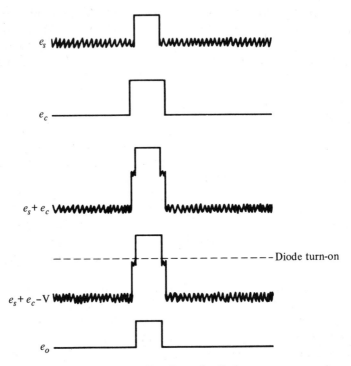

FIGURE 4-10. Waveforms for diode gate.

the control signal can be designed for two or more controls. A summing network, the output of which controls the diode in Fig. 4-9, is all that is required. In this case, the signal voltage will not appear at the output of the gate unless all the required control signals arrive at the gate at the same time.

4-8 THE BIDIRECTIONAL GATE

Transistors and vacuum tubes are commonly used in transmission gates since they can easily handle a bidirectional signal. Several typical curcuits are shown in Fig. 4-11. All of these circuits have the advantage of being bidirectional—that is, the transmitted signal can be of either polarity or alternating polarity and still be transmitted. The reason for the bidirectional feature is obvious. The control signal will be applied in a manner so as to bring the transistor or tube out of cutoff and into the active region. Once biased in the active region, the transistor or tube operates as a normal amplifier until the control signal is removed.

The Bidirectional Gate

The transistor and triode circuits of Fig. 4-11 are identical in that the control is applied to the same terminal as the signal. The pentode circuit can also be operated in this manner; however, it is more flexible in that it can be

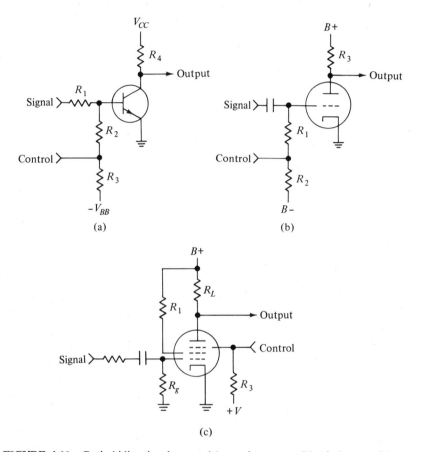

FIGURE 4-11. Basic bidirectional gates: (a) transistor gate; (b) triode gate; (c) pentode gate.

cut off at either of the two remaining grids. Coincidence operation is also possible where two or three of the grids accept control gates that must be coincident before the signal is transmitted. In the pentode circuit illustrated, the control gate is applied to the suppressor grid to override the effect of a negative supply at that point.

4-9 ELIMINATION OF THE PEDESTAL

All of the circuits in Fig. 4-11, though providing the advantages of gain and bidirectional operation, have a disadvantage in that the output signal rides on a pedestal. A common technique used to remove the pedestal employs a second transistor or a tube in parallel and drawing current through the same load resistor. A circuit is shown in Fig. 4-12. In the absence of a gate signal, T_2 is biased in the active region and T_1 is cut off. The gate signal applied negative to T_2 and positive to T_1 cuts off T_2 and places T_1 at the same point in the active region where T_2 was before the gate was applied. During the gate time, T_1 is prepared to transmit the signal to the output. Because each transistor draws the same amount of current through R when it is not cut off, the output voltage level is held constant, with the signal superimposed on that level.

4-10 A BIDIRECTIONAL DIODE GATE

A bidirectional diode gate is shown in Fig. 4-13. Commonly called a six-diode gate, it features, along with bidirectionality, minimum signal attenuation, and no pedestal. In the absence of a gate, diodes D_1 and D_2 are maintained in conduction by the negative and positive control voltages, respectively. Thus, point A is clamped to a negative potential, point B to a positive potential, and D_3, D_4, D_5, and D_6 are reverse biased. There is no output voltage. The gate consists of a reversal of polarity of each control as shown in the waveform of Fig. 4-13. During the gate period, D_1 and D_2 are reverse biased and the other four diodes conduct with electron flow from $-V$ to $+V$ through both sides of the bridgelike diode array. Regulated voltages V and $-V$, along with precision resistors R and matched diodes, insure that there is no pedestal at the output. The gate amplitude is not critical, so long as D_1 and D_2 are cut off. Points A and B are nearly at ground (0 V) potential due to the balance of the circuit during the gate period. A positive signal applied during the gate period will cut off D_3 and clamp point B to the signal amplitude, which cuts off D_6. Current flows through D_4 and R, developing a positive output signal, which is limited in amplitude by the potential at point B and therefore represents a good reproduction of the input signal. For a negative input signal, point A is clamped to the signal level and current flows to the $-V$ supply through D_6 and R. The overall effect is that when the gate is applied, the conducting diodes represent a low-resistance path from signal source to R_L.

A Bidirectional Diode Gate

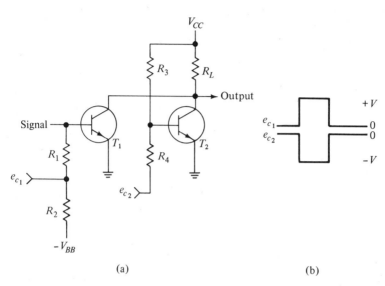

FIGURE 4-12. (a) Circuit for elimination of pedestal; (b) waveforms.

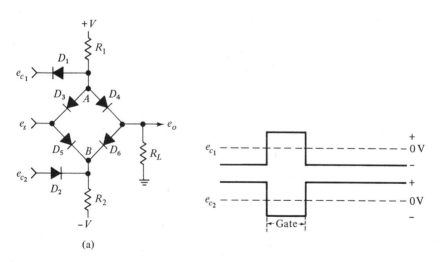

FIGURE 4-13. Bidirectional diode gate.

4-11 NONIDEAL CONDITIONS

Throughout our discussion of transmission gates we have considered only the ideal circuits. For most applications, the assumption of ideal conditions is sufficient to accurately determine conditions in the circuit at any time. In those cases where high switching rates with narrow (fast rise time) gates are required, more accurate results can be obtained only if all known capacitances and resistances are included in the circuit calculation. The discussion of diode and transistor response in Chapters 2 and 3 applies in all cases; however, the easiest solution to a problem frequently is obtained in the laboratory, beginning with a circuit based on calculations using the ideal assumptions.

Review Questions

1. Define the following logic gates: (a) OR, (b) NAND, (c) AND, (d) NOR.
2. Explain the purpose of a truth table.
3. Define *positive logic* and *negative logic* and state an example of each.
4. Explain the operation of the NOT circuit in Fig. 4-7.
5. State at least five types of logic circuits and give an example of each.
6. Explain the difference between a transmission gate and a logic gate.
7. What is the purpose of a control signal or timing signal as applied to gate circuits?
8. Explain the operation of the circuit of Fig. 4-9.
9. What is the function of the control voltage in the bidirectional gate circuit of Fig. 4-11?
10. What is an obvious disadvantage of the bidirectional gates illustrated in Figs. 4-11(a), (b), and (c)?
11. Explain the operation of the bidirectional diode gate of Fig. 4-13 when a negative input signal (e_s) is applied.

Problems

1. Draw a logical diagram of a NAND (NOT-AND) circuit and state the associated truth table for A and B inputs. (Hint: study Fig. 4-4.)
2. Repeat Problem 1 for the NOR gate circuit.
3. Redraw the circuit of Fig. 4-6 for negative logic and define the zero and one levels.
4. What is the current in diode D_2 of Fig. 4-5 if R is 5 kΩ and A, B, and C inputs are 0, 0, and 1, respectively? What is the current in R?

Problems

5. Redraw the circuit of Fig. 4-5 to operate with negative logic and state the associated truth table, using voltage levels.
6. Assume the circuit of Fig. 4-9 to have the following circuit parameters and solve for the indicated quantities: $R_1 = 10$ kΩ, $R_2 = 3$ kΩ, $V = -10$ V, $R = 20$ kΩ, $e_C = +4$ V, $e_s = 0$ V.
   Find: (a) e_o
   (b) I_1
   (c) I_2
7. Solve the circuit of Example 4-4 for the same indicated quantities if R_1 is 15 kΩ, and V_{in} switches from 0 V to $+10$ V. All other values remain the same.
8. Solve the circuit of Fig. 4-6 for the stated quantities if the ground end of R is connected to a -6 V supply and inputs A and B are $+3$ and $+8$, respectively. $R = 7$ kΩ.
   Find: (a) e_{out}
   (b) I_R
   (c) I_1, I_2
9. Draw a schematic diagram of a negative OR gate with four inputs and explain the operation of the circuit.
10. Consider the circuit of Fig. 4-11(a) and analyze the circuit with the stated circuit parameters: $R_1 = 10$ kΩ, $R_2 = 5$ kΩ, $R_3 = 20$ kΩ, $R_4 = 4$ kΩ, $V_{CC} = +12$ V, $V_{BB} = -10$ V, $V_{in} = +5$ V, $V_C = 0$ V. Assume that the transistor is an ideal device operating in the "forced β" mode.
    Find: (a) $I_{B(on)}$
    (b) I_2
    (c) I_3
    (d) I_4
    (e) V_{out}
11. Solve the circuit of Problem 10 for the given quantities if V_C is -5 V. All other circuit values remain the same. Also solve the circuit for a V_C of $+5$ V.
12. Determine the following quantities of the circuit in Fig. 4-13 for the given circuit values: $V = \pm 10$ V, $R_1 = 5$ kΩ, $R_2 = 5$ kΩ, $R_L = 7.5$ kΩ; all diodes are considered ideal. The voltage level for e_{C_1} and e_{C_2} are ± 4 V, and e_s is $+6$ V.
    Find: (a) V_{out}
    (b) I_1
    (c) I_2
13. What effect would an open D_5 have on the operation of the circuit in Problem 12?

Multivibrators

5

In this chapter attention is given only to multivibrators. These regenerative circuits enjoy widespread employment in virtually every branch of electronics.

The bistable multivibrator is a two-state circuit that requires a trigger input to change from one state to the other. The next trigger input will cause the circuit to revert to its original state. Because the circuit remains in a given state determined by the last trigger received, it is often referred to (and used as) a memory circuit.

The monostable multivibrator has a single stable state. When it receives a trigger input the circuit switches to a quasistable state for a given period of time, and then resets itself to the stable state. It is used extensively as a gate generator.

The astable multivibrator is simply a relaxation oscillator. It requires no trigger input and provides a square wave output. The circuit can be synchronized by a stable oscillator and used as a basic timing pulse generator.

Each of the types of multivibrator mentioned above can be designed with either vacuum tubes or transistors. However, the current trend is toward transistors. Because of this, we will direct our attention mainly to the transistorized versions of each circuit, emphasizing at this point that the vacuum-tube counterparts are identical in operation.

5-1 THE BISTABLE MULTIVIBRATOR (FLIP FLOP)

The flip flop is one of the most common circuits employed in digital computer technology, but it has applications in many other areas as well. Among the more common names for the circuit are the Eccles-Jordan (after the in-

The Bistable Multivibrator (Flip Flop)

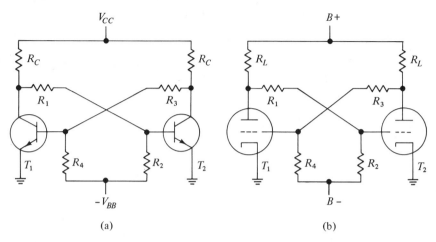

FIGURE 5-1. (a) The transistor flip flop; (b) vacuum tube flip flop.

ventors), binary, and scale-of-two circuit. Both vacuum-tube and transistor versions of a typical circuit are shown in Fig. 5-1. The circuit is a regenerative one with two stable states. Whenever one transistor (tube) is at cutoff, the other is conducting heavily. An input pulse (trigger) is required to change the state of the circuit, while a second trigger simply causes it to revert to its original state.

Consider the transistor circuit of Fig. 5-1 and assume that power has been applied and a stable state achieved. Thus, T_1 might be on (conducting heavily) and T_2 cut off, in which case the collector potential for T_1 would be low. Under this condition, the voltage divider R_1 and R_2 would establish a negative potential at the base of T_2, back biasing that transistor and insuring cutoff. Thus T_1 is forward biased, as we assumed initially, and the circuit is in a stable state.

Now suppose a negative trigger is applied to the base of T_1. If the amplitude is sufficient, T_1 will be cut off and the collector voltage will rise sharply. The $R_1 R_2$ divider transmits this voltage change to the base of T_2, and that transistor becomes forward biased. The collector potential of T_2 drops, and now T_1 is cut off and remains that way, even though the original trigger is removed.

Qualitative analysis of the vacuum-tube circuit of Fig. 5-1 is identical. Furthermore, in both circuits the initial stable state is determined in a random manner because of variations in components. That is, when power is applied initially, one transistor (tube) will conduct more than the other for any of a number of possible reasons. Because the circuit is regenerative, this initial imbalance is sufficient to drive the circuit to a stable state.

5-2 TRIGGERING THE FLIP FLOP

The basic circuits illustrated in Fig. 5-1, although practical, have a minor drawback in that two separate trigger inputs are required, one to each base (grid). A more common requirement for the flip flop is that it must react to every trigger pulse that appears at the output of a particular circuit. In other words, successive triggers on the same line must cause the flip flop to change state, once for every trigger.

One simple technique that is used is illustrated in Fig. 5-2 (which also

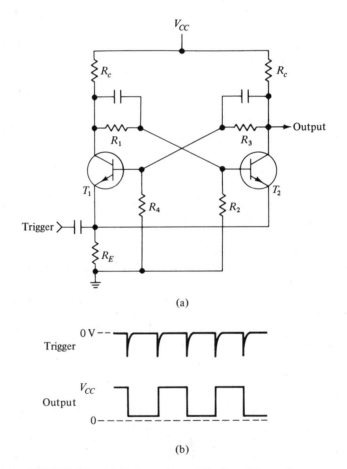

FIGURE 5-2. (a) Emitter triggered flip flop; (b) waveforms.

Triggering the Flip Flop

shows the output waveform from one collector). In this circuit, the common emitter resistor R_E serves as an input to both sides of the flip flop. (It also provides self-bias, eliminating the need for a second power supply.) Negative triggers applied at the common emitter junction will affect only the transistor, which is cut off.

The coupling capacitors shown in Fig. 5-2 are commonly used to speed up the switching process, as they effectively short-circuit the coupling resistors for instantaneous changes in collector voltage. Thus, at the instant the trigger is applied, the resultant fluctuation in collector potential is instantaneously transmitted to the base of the opposite transistor.

The emitter triggered circuit in Fig. 5-2(b), although simple, requires a large trigger current. In addition, it will retrigger on the trailing edge of a pulse if the pulse duration is longer than the circuit transient period. That is, the circuit is sensitive to both positive and negative changes in trigger potential.

The sensitivity of a flip flop to bipolar triggers may or may not be a desirable feature. Generally, unipolar triggers are used in conjunction with diode steering circuits that direct the trigger to the proper transistor. One good reason for diode steering is that quicker response is achieved by applying a trigger to the conducting transistor rather than to the one that is cut off. This peculiarity is easy to understand when one considers that the conducting transistor (especially when it is not saturated) is in the active state. Thus, any change at its input results in an immediate fluctuation at the output. The same is not true for the cutoff transistor. Before any response occurs at its collector, the trigger must rise above (or fall below) the turn-on voltage—that is, it must overcome the cutoff bias.

Two circuits that employ diode steering are illustrated in Fig. 5-3. The flip flop of Fig. 5-3(a) uses a steering circuit to direct negative triggers to the base of the conducting transistor. Suppose, for example, T_2 is conducting and T_1 is cut off. Then the high potential at the base of T_2 places D_2 in forward bias. D_1 is reverse biased since the base of T_1 is at a lower potential than the base of T_2. Because of the bias condition of the diode, a negative trigger will be steered to T_2 and initiate the change-of-state operation. When the circuit flips, the conditions will be exactly reversed, with D_1 conducting and D_2 reverse biased. Thus, the next trigger will be steered to T_1.

The circuit of Fig. 5-3(b) is similar, except that the steering network is in the collector circuit. The steering circuit itself is identical, the diodes being reversed only to accommodate the positive triggers required by the PNP transistors. As should be expected, collector triggering requires larger pulses than base triggering.

Steering circuits, speed-up capacitors, and self-bias as discussed in the pre-

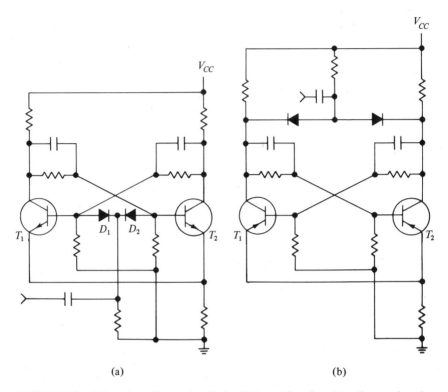

FIGURE 5-3. Triggering with steering diodes: (a) base triggering; (b) collector triggering.

ceding paragraphs are common to vacuum-tube circuits as well as to those that employ transistors. The terminology used would be consistent with the device, and terms such as plate triggering, grid triggering, and cathode coupling are common in vacuum-tube applications.

5-3 SATURATING VS NONSATURATING FLIP FLOPS

Thus far, the circuits we have discussed have been of the saturating variety. That is, when a transistor is on, it operates in the saturation portion of the characteristic curves. There is a singular disadvantage associated with a saturating flip flop. Due to the minority carrier storage effect in the base of a saturated transistor, the response time is not as fast as it could possibly be. By preventing the transistor from entering saturation, an improvement in response is achieved. A number of techniques are available to insure that a

Cathode Coupled Binary (Schmitt Trigger)

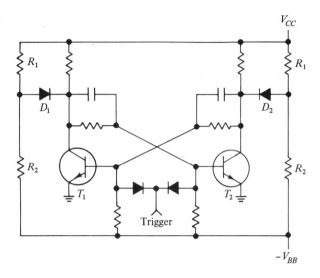

FIGURE 5-4. Nonsaturating flip flop.

transistor does not saturate, the most common of which employs diode clamping at the collector. Such a circuit is shown in Fig. 5-4. Diodes D_1 and D_2 couple the collectors to a reference voltage level formed by the R_1R_2 dividers. The reference level is set slightly above the saturation voltage for the transistor. Before the transistor can saturate, the diode clamps the collector to the reference voltage.

The improved response time obtained with the nonsaturating flip flop is gained at a considerable expense in terms of power consumption. This disadvantage must be weighed against the saturating flip flop, where little current is drawn through the cutoff transistor and only a small voltage is developed across the saturated one.

5-4 CATHODE COUPLED BINARY (SCHMITT TRIGGER)

An important binary which is slightly different from the one previously discussed is shown in Fig. 5-5. (To emphasize the fact that differences between vacuum-tube and transistor multivibrators are minor, we will discuss the vacuum-tube version of the Schmitt trigger.) The circuit employs only one cross-coupling resistor, from T_1 to the grid of T_2. Feedback from T_2 to T_1 is accomplished through the common cathode (emitter) resistor.

Assume that the circuit is at rest, with T_1 cut off and T_2 conducting (not

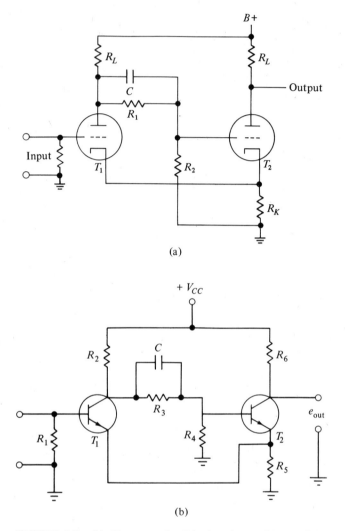

FIGURE 5-5. (a) Vacuum tube Schmitt trigger; (b) transistor Schmitt trigger.

necessarily saturated). The conduction state of T_2 is determined by the voltage difference between its grid and cathode (base and emitter). The cathode (emitter) potential is determined by the amount of current that T_2 draws through R_k (R_E); the grid (base) potential is a result of the voltage divider

Cathode Coupled Binary (Schmitt Trigger)

action of R_1 and R_2 in conjunction with the plate voltage (collector) of T_1. The cutoff state of T_1 is due simply to the large potential developed across the common cathode (emitter) resistor by T_2.

Application of a positive voltage to the grid of T_1 results in an increase in plate current and a decrease in the plate voltage that is divided and applied to the grid of T_2. The trigger pulse must be of sufficient amplitude to bring T_1 out of cutoff, which in tube circuits may be rather large (50 V or more). Once conduction begins, however, a relatively small potential change at the plate of T_1 will cause cutoff of T_2. Cutoff of T_2 is insured only if the potential at the grid of T_2 results in an input (via the cathode resistor) to T_1, which, when amplified by T_1, is sufficient to maintain the cutoff of T_2. In other words, the feedback loop gain must be greater than unity. An increase in the potential at the grid of T_1 will have no further effect on the circuit once triggering has occurred.

Removal of the input voltage will result in retriggering of the circuit to its original state. It is extremely important to note, however, that a decrease in potential to the original triggering value will not retrigger the circuit. That is, for positive-going signals one input voltage causes the circuit to trigger, and for negative-going signals a different input voltage causes triggering. This is what is commonly called the *hysteresis* characteristic of the circuit.

The reason for the hysteresis effect is easy to understand. The potential required to bring about the cutoff condition of T_2 is approximately that required to bring T_1 out of cutoff, as we have just seen. (In other words, a very small grid-cathode potential difference exists when T_2 is in the conduction state.) A much larger potential difference between the grid and cathode of T_2 is developed when that tube is in cutoff. As a result, a greater plate voltage swing is required of T_1 to put T_2 back into conduction. The effect is illustrated quite clearly in Fig. 5-6, which also shows one of the more popular usages of the circuit—that is, the squaring of nonsymmetrical waveforms.

The transistorized version of the Schmitt trigger is shown in Fig. 5-5(b).

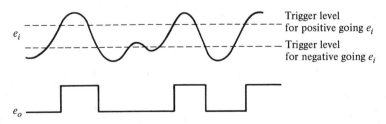

FIGURE 5-6. Squaring of nonsymmetrical waveform by Schmitt trigger.

This circuit is sometimes called an emitter coupled flip flop, and the principle of operation is identical to that of the tube circuit just discussed.

5-5 THE MONOSTABLE MULTIVIBRATOR (ONE SHOT)

As the name indicates, the monostable multivibrator has a single stable state. When a trigger pulse is applied, the circuit is switched to an unstable, or quasistable, state. For a given time following the application of the trigger, the circuit remains in the quasistable state, and then it reverts to the original state. Since the duration of the quasistable state is determined by design, the circuit is useful for the generation of gates and delayed triggers.

A typical design for the one shot is shown in Fig. 5-7. The circuit is nearly identical to the flip flop except that one of the cross-coupling paths is removed, which prevents T_1 from maintaining cutoff potential at the base of T_2. With the initial application of power, the circuit goes to the stable state. For a typical transistor the circuit conditions might be

for T_1, $V_{CE(\text{off})} = V_{CC}$, $V_{BE(\text{off})} = -.3$ V, $V_e = 0$
for T_2, $V_{CE(\text{sat})} \approx 0$, $V_{BE(\text{sat})} = .5$ V, $V_e = 0$, $V_{BE(\text{on})} = .4$ V.

A negative trigger applied to the collector of T_1 is coupled via C to the base of T_2 and cuts it off, the operation being identical to that in switching the state of a flip flop. Because there is no resistive coupling from the collector

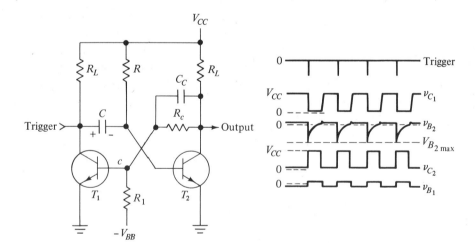

FIGURE 5-7. The one-shot multivibrator and waveforms.

The Monostable Multivibrator (One Shot)

of T_1 to the base of T_2, the quasistable state is maintained for as long as the base of T_2 remains at cutoff. The potential at the base of T_2 remains below cutoff only until capacitor C charges to the turn-on voltage for the transistor. The waveforms of Fig. 5-7 indicate the operation when a trigger is applied. Assuming that the action of the trigger causes T_1 to be driven into saturation, the change in potential at the base of T_2 will equal the change in collector voltage of T_1, or $-V_{CC}$. C then charges from $-V_{CC}$ toward $+V_{CC}$. When T_2 turns on, the potential on C is clamped to the base saturation voltage, which for most transistors is on the order of .5 V. With the turn-on of T_2 the cycle is completed, and the circuit awaits the next trigger.

The duration of the output pulse (gate) generated is determined by the RC time constant V_{CC} and the turn-on potential for the transistor. The simplified circuit of Fig. 5-8(a) is used to calculate the potential at the base of T_2 and, ultimately, the gate width. Prior to the application of the trigger, the voltage $v_{B_2(\text{sat})}$ is the saturation voltage for T_2 and is constant for a given transistor and circuit configuration. At $t = 0+$ (immediately after the trigger), the voltage across the capacitor must remain unchanged; therefore, the instantaneous drop in potential at point B when T_1 is on (saturated) is felt also at point A, and

$$v_A = v_{B_2(\text{sat})} - V_{CC}. \qquad (5\text{-}1)$$

Immediately, point A will begin to charge toward V_{CC} through R. The time constant Eq. (1-8) applies, and

$$\begin{aligned} v_A &= V_{CC} - [V_{CC} - (v_{B_2(\text{sat})} - V_{CC})]e^{-t/RC} \\ &= V_{CC} - (2V_{CC} - v_{B_2(\text{sat})})e^{-t/RC}. \end{aligned} \qquad (5\text{-}2)$$

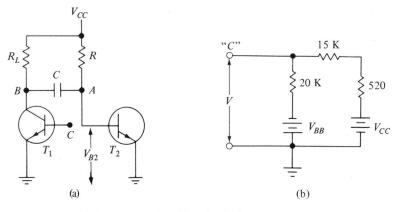

FIGURE 5-8. (a) Simplified circuit; (b) equivalent circuit.

But v_A can charge only until T_2 turns on, at which time the circuit reverts to its stable state. Thus, for a given turn-on voltage ($v_{B_2(\text{on})}$), the gate duration is found from

$$e^{-t/RC} = -\frac{v_{B_2(\text{on})} - V_{CC}}{2V_{CC} - v_{B_2(\text{sat})}}, \tag{5-3}$$

so that

$$t = RC \ln \frac{2V_{CC} - v_{B_2(\text{sat})}}{V_{CC} - v_{B_2(\text{on})}} \tag{5-4}$$

$$\approx RC \ln 2 = .69RC. \tag{5-4a}$$

Equation (5-4a) provides a reasonable approximation for the gate duration in a monostable circuit. The assumptions made to arrive at the equation are:

1. $v_{C_1} = 0$, when T_1 is in saturation.
2. $v_{B_2(\text{sat})} \ll V_{CC}$.
3. $v_{B_2(\text{on})} \ll V_{CC}$.

The small overshoot that occurs at the base of T_2 after turn-on is generally negligible; however, the circuit should not be retriggered during this time. Any trigger that arrives before the circuit reaches its final stable state will result in erratic operation due to an inconsistent initial charge on C.

Example 5-1 Determine the stated quantities of the circuit in Fig. 5-7 for circuit parameters of $R_L = 520 \; \Omega$, $R_C = 15 \; \text{k}\Omega$, $R_1 = 20 \; \text{k}\Omega$, $R = 10 \; \text{k}\Omega$, $V_{CC} = +15$, $C = .01 \; \mu\text{F}$, $C_C = 160 \; \text{pF}$, and $V_{BB} = -8 \; \text{V}$. $T_1 = T_2$ and $V_{CE_1(\text{sat})} = V_{CE_2(\text{sat})}$.

Find: (a) gate duration t
 (b) maximum turn-off ($V_{B_2(\text{off})}$) T_2
 (c) Prove that T_1 will turn on when the circuit is triggered.

Solution:
(a) Solving for t from Eq. (5-4a),

$$t = RC \ln 2 = .69RC$$
$$= (10^4)(10^{-8})(.69) = (.69)(10^{-4}) = 69 \; \mu\text{sec}.$$

Therefore, the gate length is 69 μsec.

(b) At the time the circuit is triggered T_2 is turned off, and the collector voltage of T_2 rises toward V_{CC} and is clamped at a predetermined voltage provided by the voltage divider network of R_L, R_C, and R_1. This rise in collector voltage of T_2 is partially coupled to the base of T_1, causing T_1 to saturate. The voltage at the base of T_2 is the sum of the capacitor C voltage and the saturation voltage of T_1. Therefore,

$$V_{B_2(\text{off})} = -V_{CC}.$$

Assuming that $V_{CE_1(\text{sat})}$ is negligible compared to V_{CC}, then

$$V_{B_2(\text{off})} \text{ maximum} = -V_{CC} = -15.$$

This magnitude of voltage accounts for the large negative spike on the waveform V_{B_2} of Fig. 5-7.

(c) Assuming that the trigger pulse is sufficient to drive T_2 off, the equivalent circuit to prove that T_1 will turn on is shown in Fig. 5-8(b). The voltage point C to ground must be positive to insure turn-on of T_1. Therefore,

$$V(\text{point } B) = IR_1 - V_{BB} = \frac{(V_{CC} + V_{BB})}{R_1 + R_C + R_L} R_1 - V_{BB}$$

$$= 12.9 - 8 = 4.9 \text{ V}.$$

This voltage (4.9 V) is not actually present at point C because the base emitter junction is connected and will change this voltage to a diode drop. However, this analysis demonstrates that T_1 will conduct under the stated conditions, for point C is positive in relation to the emitter of T_1.

5-6 THE VACUUM-TUBE ONE SHOT

As with the flip flop, the vacuum-tube counterpart of the one shot is identical in operation to the transistor version. A typical design is shown in Fig. 5-9. The stable condition is such that T_2 is normally conducting and T_1 is in cutoff. Under these conditions the grid of T_2 will draw current (clamping the voltage on C) and the plate voltage will be low. The $B-$ supply maintains

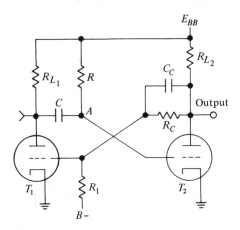

FIGURE 5-9. Vacuum tube one-shot multivibrator.

T_1 solidly in cutoff until a trigger is applied. When the transition occurs, if the swing at the plate of T_2 is sufficient, the grid of T_1 may be driven above the minimum turn-on voltage, where it may draw current. In any event, whenever a conducting grid is encountered, it is not unreasonable to assume a resistance value on the order of 1 kΩ. In other words, the grid cathode of a saturated tube forms a clamping diode.

To determine the gate duration, one applies the same reasoning as was used in the transistor version. Thus,

$$t = RC \ln \frac{2E_{BB} - E_{b_1(\text{sat})}}{E_{BB} - V_{g_2(\text{on})}}$$

$$V_{g_2(\text{on})} = E_{BB} - (E_{BB} + I_1 R_L)e^{-t/RC}$$

and

$$I_1 R_L = E_{BB} - E_{b_1(\text{sat})}.$$

Therefore,

$$V_{g_2(\text{on})} = E_{BB} - (E_{BB} + E_{BB} - E_{b_1(\text{sat})})e^{-t/RC}$$
$$= E_{BB} - (2E_{BB} - E_{b_1(\text{sat})})e^{-t/RC}.$$

Solving for t,

$$t = RC \ln \frac{2E_{BB} - E_{b_1(\text{sat})}}{E_{BB} - V_{g_2(\text{on})}}. \tag{5-5}$$

$E_{b_1(\text{sat})}$ and $V_{g_2(\text{on})}$ can easily be obtained from a load line plot on a plate characteristic curve.

Example 5-2 Determine the gate duration of the vacuum-tube one-shot multivibrator (Fig. 5-9) for the given circuit parameters:

$R_{L_1} = R_{L_2} = 10$ kΩ, $R = 100$ kΩ, $C = .2$ μF, $R_C = 33$ kΩ,
$R_1 = 68$ kΩ, $B = -125$ V, $E_{BB} = 250$ V, $T_1 = T_2$,
$E_{b_1(\text{sat})} = 60$ V, and $V_{g_2(\text{on})} = -12$ V.

Solution: Eq. (5-5) is

$$t = RC \ln \frac{2E_{BB} - E_{b_1(\text{sat})}}{E_{BB} - V_{g_2(\text{on})}}$$

$$= 10^5(2 \cdot 10^{-9}) = \ln \frac{2(250) - 60}{250 + 12}$$

$$= (2 \cdot 10^{-4}) \ln (1.67) = 102.4 \text{ μsec}.$$

5-7 EMITTER COUPLED ONE SHOT

A popular version of the one shot is the emitter coupled circuit shown in Fig. 5-10. In the stable state, T_2 is saturated (or biased in the upper active region) and draws current through the common emitter resistor, developing

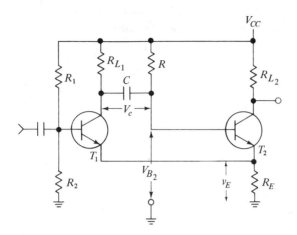

FIGURE 5-10. Emitter coupled one-shot multivibrator.

a positive emitter potential for both transistors. This potential, combined with the voltage divider action of R_1 and R_2, causes the cutoff of T_1. To insure cutoff, then, V_E must be greater than $[R_2/(R_1 + R_2)]V_{CC}$. A positive trigger applied to the base of T_1 results in conduction and a drop in collector voltage. C transmits the negative-going potential to the base of T_2 and reduces its conduction state to cutoff. T_1 is maintained in conduction by the potential developed across R_2, which is now greater than V_E since the current draw through the emitter resistor is greatly reduced by the cutoff of T_2. Following the cutoff of T_2, C begins to charge toward V_{CC} through R, and, eventually, T_2 turns on to restore the stable state.

Prior to the application of the trigger, the potential at the base of T_2 is

$$V_{B_2} = V_E + V_{BE(\text{sat})} = V_{B_2}(t = 0-) \tag{5-6}$$

and

$$V_C = V_{CC} - V_{BE_2(\text{sat})} - V_E. \tag{5-6a}$$

Immediately after the trigger is applied,

$$V_{B_2} = -V_C - I_{C_1}R_{L_1} + V_{CC}. \tag{5-7}$$

Substituting Eq. (5-6a) into Eq. (5-7) yields

$$V_{B_2} = V_{BE(\text{sat})} + V_E - I_{C_1}R_{L_1}$$

and

$$V_{B_2} = V_{B_2}(t = 0-) - I_{C_1}R_{L_1}, \tag{5-7a}$$

which, for the determination of gate width, becomes the initial voltage on C.

The capacitor must now charge from this initial voltage toward V_{CC}. Again, the basic time constant equation applies and

$$V_{B_2} = V_{CC} - (V_{CC} - [V_{B_2}(t = 0-) - i_{C_1}R_{L_1}])e^{-t/RC}. \quad (5\text{-}8)$$

For a given circuit and transistor, a specific value of v_{B_2} exists for the turn-on of T_2. This value is the normal potential required to forward bias the emitter base junction plus the emitter voltage established by the conduction of T:

$$V_{B_2(\text{on})} = V_E + V_{BE(\text{on})} \approx V_E. \quad (5\text{-}9)$$

Substitution of this voltage into the preceding equation results in the equation that is used to determine the gate width for the emitter coupled one shot:

$$t = RC \ln \left[\frac{V_{CC} - V_{B_2}(t = 0-) + I_{C_1}R_{L_1}}{V_{CC} - V_E} \right]. \quad (5\text{-}10)$$

Notice that

$$I_{C_1}R_{L_1} = V_{CC} - V_{CE_1(\text{on})} - V_E. \quad (5\text{-}10a)$$

$V_{CE_1(\text{on})}$ is the collector voltage of T_1 when a trigger pulse has turned T_1 on. This value is often obtained from a load line plot on the transistor characteristic curve. Substituting Eq. (5-6) and Eq. (5-10a) into Eq. (5-10) gives

$$t = RC \ln \frac{2(V_{CC} - V_E) - (V_{BE(\text{sat})} + V_{CE_1(\text{on})})}{V_{CC} - V_E}. \quad (5\text{-}11)$$

If T_1 is allowed to go into saturation and not into the active region, $V_{CE_1(\text{on})} = V_{CE(\text{sat})}$ and Eq. (5-11) may be approximated by Eq. (5-4a).

Note that the gate duration for the emitter coupled circuit is a function of the bias voltage established by the R_1R_2 divider. This is simply because V_E and I_CR_L are both determined by the bias potential. Replacement of the fixed divider with a potentiometer provides a circuit with a variable delay and, as such, is extremely popular.

The vacuum-tube counterpart of the emitter coupled one shot is the cathode coupled circuit shown in Fig. 5-11. Circuit operation is nearly identical. The development of the circuit equation is left as an exercise.

Example 5-3 Determine the gate duration of the circuit in Fig. 5-10. The circuit values are: $R_1 = 47$ kΩ, $R_2 = 10$ kΩ, $R_{L_1} = R_{L_2} = 1$ kΩ, $R_E = 470$ Ω, $C = 1000$ pF, $R = 68$ kΩ, $V_{CC} = +14$ V, $V_E = +4$ V, $V_{CE_1(\text{on})} = +2$ V, $V_{BE(\text{sat})} = .6$ V.

Solution: The gate duration is given by Eq. (5-11). Therefore,

$$t = \tau \ln \frac{2(V_{CC} - V_E) - (V_{BE(\text{sat})} + V_{CE_1(\text{on})})}{V_{CC} - V_E}.$$

The Astable Multivibrator

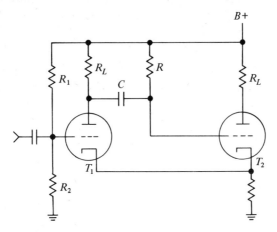

FIGURE 5-11. Cathode coupled multivibrator.

Now

$$\tau = RC$$
$$= (68 10^3)(10^3)(10^{-12}) = 68 \ \mu\text{sec}$$

and

$$t = 68 10^{-6} \ln \frac{2(14-4) - (.6 + 2)}{(14-4)}$$
$$= 37.4 \ \mu\text{sec}.$$

5-8 THE ASTABLE MULTIVIBRATOR

The astable circuit, sometimes called a free-running multivibrator, is shown in Fig. 5-12. Each transistor (tube) operates in a quasistable state, which is similar to that state described for the one-shot circuit. Capacitive cross-coupling from the collector (plate) of each transistor (tube) to the other is characteristic of the collector (plate) coupled circuit.

Circuit operation is nearly identical to that of the flip flop, except that no trigger is required. Initial application of power will cause the circuit to operate—as a larger current will be developed in one of the transistors. This results in a negative-going potential at the collector, which is cross-coupled to the base of the other transistor.

Assume the circuit is in operation and T_1 is off while T_2 is saturated. The waveforms of Fig. 5-13 apply. At the instant T_2 saturates, the drop in potential at the base of T_1 is

$$I_{C_2} R_{L_2} \approx V_{CC}. \tag{5-12}$$

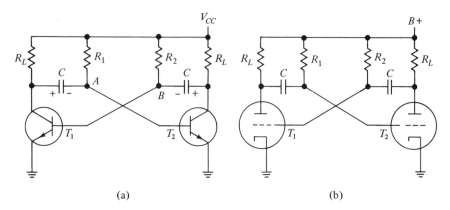

FIGURE 5-12. (a) Transistor free-running multivibrator; (b) vacuum tube free-running multivibrator.

Point B in Fig. 5-12(a), therefore, will begin charging from $-V_{CC}$ to V_{CC}. When the turn-on potential for T_1 is reached, a drop $I_{C_1}R_{L_1}$ will be realized at the collector of T_1 and the base of T_2. Thus, T_2 cuts off and T_1 is driven to saturation. Point A goes through an identical excursion to that described for

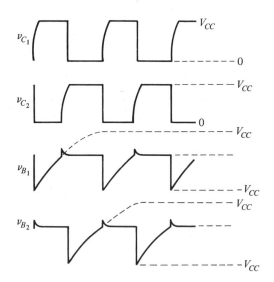

FIGURE 5-13. Waveforms for the free-running multivibrator.

The Astable Multivibrator

point B to complete a cycle. Because there are no resistive components in the cross-coupling branches, neither transistor can be maintained in a stable state.

Assuming that the turn-on voltage for the transistors is small with respect to V_{CC}, and, further, that the transistors operate between saturation and cutoff, the equation for the off period of each transistor is

$$t = RC \ln 2$$
$$= .69RC. \tag{5-13}$$

For a symmetrical circuit, t represents the half period of the output square wave, which can be taken from either collector. Nonsymmetrical outputs can be achieved simply by adjustment of the RC values associated with each transistor. Rounding of the waveform when the transistor is turned on is due to the storage time delay associated with bringing a transistor out of saturation. Diode clamping at the collectors can be used to prevent the transistors from saturating, thereby yielding a better square wave output. The price paid for the improvement in waveshape is a higher average power consumption per cycle of operation—the same disadvantage we encountered in the case of the nonsaturating flip flop.

Example 5-4 What is the frequency of the astable multivibrator illustrated in Fig. 5-12 if the circuit values are: $C_1 = .002\ \mu\text{F}$, $C_2 = 910\ \text{pF}$, $R_1 = 12\ \text{k}\Omega$, $R_2 = 20\ \text{k}\Omega$, $V_{CC} = +12\ \text{V}$, $T_1 = T_2$?

Solution: Observing that the RC values are different for each transistor (therefore, the multivibrator is nonsymmetrical), the frequency is given by

$$F = \frac{1}{t} = \frac{1}{t_1 + t_2}.$$

Equation (5-13) is
$$t = RC \ln 2.$$
Also,
$$t_1 = R_1 C_1 \ln 2$$
and
$$t_2 = R_2 C_2 \ln 2$$
$$t_1 = 16.6 \times 10^{-6}\ \text{sec}$$
$$t_2 = 12.55 \times 10^{-6}\ \text{sec}.$$
Thus,
$$t = t_1 + t_2 = 29.15\ \mu\text{sec}$$
and
$$F = \frac{1}{29.15 \times 10^{-6}} = 34.3\ \text{kHz}.$$

Review Questions

1. What is meant by the term *quasistable state*?
2. Define each of the following multivibrators: (a) astable, (b) bistable, (c) monostable.
3. What is another name often used for each of the multivibrators discussed in Question 2?
4. Discuss three methods used to trigger a flip flop.
5. Describe the fundamental difference between a saturating and a nonsaturating bistable multivibrator.
6. What are the factors that affect the gate length of a one-shot multivibrator?
7. State at least one disadvantage and one advantage of utilizing a nonsaturating flip flop as compared to a saturating flip flop.
8. What would be the effect on gate duration of the circuit in Fig. 5-10 if R_E were increased in value?
9. Explain the meaning of a *nonsymmetrical output* of an astable multivibrator.
10. What is the meaning of *hysteresis* in reference to a Schmitt trigger?
11. Explain the unique feature that characterizes an emitter coupled binary circuit.
12. Explain how diode clamping may prevent a transistor from saturating.
13. Discuss the meaning of $V_{BE(on)}$ and $V_{BE(sat)}$ in reference to transistors used in multivibrators.
14. What is the relative input resistance of a vacuum tube when drawing grid current? Discuss this action.
15. Refer to Fig. 5-3(b) and explain what takes place in the diode steering circuit when a pulse is applied.
16. What polarity of input is needed to trigger the circuit of Fig. 5-3(a)? Discuss this action.
17. An astable multivibrator does not require a trigger pulse to become operational; explain how this action is possible.
18. Discuss the function of the "speed-up" capacitors used in bistable multivibrators.
19. What is meant by a *stable* condition of a multivibrator?
20. What are the factors that determine the frequency of a monostable multivibrator?
21. What would be the influence on frequency if C in Fig. 5-11 were increased in value?

22. Discuss the relationship between the frequency output and the trigger input of a bistable multivibrator.

Problems

1. Solve the circuit of Fig. 5-7 for the stated quantities. The circuit values are: $R_L = 1$ kΩ, $R_C = 20$ kΩ, $R_1 = 33$ kΩ, $R = 12$ kΩ, $V_{CC} = +18$ V, $V_{BB} = -9$ V, $C = .015$ μF, $C_C = 100$ pF, $T_1 = T_2$, and $V_{CE_1(\text{sat})} = V_{CE_2(\text{sat})}$.
   Find: (a) $V_{B_2(\text{off})}$
   (b) gate duration t
   (c) Prove that T_1 will turn on when triggered.
2. Draw a schematic diagram of a PNP Schmitt trigger.
3. Determine the value for C_1 and C_2 of Fig. 5-12 if the circuit parameters are: $R_1 = R_2 = 15$ kΩ, $V_{CC} = +15$ V, $T_1 = T_2$, frequency of oscillation is 25 kHz. $C_1 = C_2$.
4. Draw a schematic diagram of a PNP saturated bistable multivibrator employing base diode steering. Positive pulses are to be used for triggering.
5. Approximate the $I_{B_1(\text{on})}$ current and $V_{B_2(\text{off})}$ voltage for the circuit of Fig. 5-1(a). Assume that T_1 is off and T_2 is on. The circuit values are: $V_{CC} = 12$ V, $V_{BB} = -6$ V, $R_C = 2$ kΩ, $R_1 = R_3 = 30$ kΩ, $R_4 = R_2 = 22$ kΩ, and $V_{CE_1(\text{sat})} = V_{CE_2(\text{sat})} = .1$ V.
6. Illustrate by a schematic diagram the changes that are necessary to modify the circuit in Problem 4 to a nonsaturating multivibrator.
7. Solve for the value of C that is needed to effect a gate length of 85 μsec for the circuit in Example 5-3. All other circuit values remain the same.
8. Determine the turn-on voltage $V_{g_2(\text{on})}$ for T_2 in Fig. 5-9. The circuit values are: $R = 120$ kΩ, $C = .001$ μF, $R_C = 47$ kΩ, $R_1 = 56$ kΩ, $E_{BB} = 275$ V, $B = -100$ V, $E_{b_1(\text{sat})} = 65$ V, and gate duration is 70 μsec.
9. Refer to Fig. 5-5(b) and state the quiescent condition (before trigger condition) of V_{CE}, I_B, and V_{BE} for each transistor. Use relative terms such as high, low, off, on, etc.
10. Repeat Problem 5–9 for the circuit of Fig. 5-10.
11. Solve the circuit of Fig. 5-1(b) for the stated quantities. Assume that T_1 is off and T_2 is on; the circuit values for $R_1 = R_3$, $R_2 = R_4$, R_L, $B+$, and $B-$ are 40 kΩ, 68 kΩ, 12 kΩ, 300 V, and 150 V, respectively. $V_{CE_1(\text{sat})} = V_{CE_2(\text{sat})} = 55$ V. (Hint: a conducting grid circuit is considered to be about 1 kΩ.)
    Find: (a) $V_{g_1(\text{off})}$
    (b) $V_{g_2(\text{on})}$

The Blocking Oscillator

6

The blocking oscillator, like the multivibrator, is a relaxation-type oscillator. It is generally used where relatively short duration pulses are required and/or where large loads are likely to be encountered. It is a circuit that always employs a pulse transformer for feedback, and frequently has an output coupled from a tertiary (third) winding of the same transformer. Because of its basic simplicity, the circuit has found widespread use in all areas of electronics, ranging from computers to radar. It can be employed in both an astable (free-running) and a monostable (one-shot) mode of operation, with either a transistor or a vacuum tube. All of these aspects will be considered in this chapter.

6-1 THE IDEAL PULSE TRANSFORMER

It is obvious from the preceding section that pulse transformer characteristics play an important role in the operation of a blocking oscillator. Ideally, the transformer is considered a lossless device that provides a small voltage gain when used in a circuit similar to that in Fig. 6-1. Practically, it is a complex device with resistance, capacitance, and inductance along with nonlinear magnetic properties. It is the purpose of this section to discuss the effects of a transformer on the shape of a pulse, and to determine which characteristics can be eliminated and which must be dealt with in our circuit considerations later on. Fortunately, for pulse considerations it is generally acceptable to consider the transformer as ideal, with minor adjustments accomplished in an external equivalent circuit.

An ideal transformer is shown in Fig. 6-2. There are N_1 primary and N_2 secondary turns, but because the transformer is to be considered ideal, it is assumed that there is no resistance associated with either the primary or secondary coil. Furthermore, a high-permeability, lossless magnetic core is

The Ideal Pulse Transformer

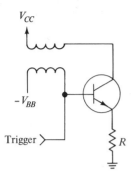
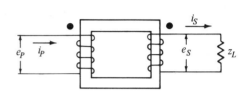

FIGURE 6-1. A monostable blocking oscillator.

FIGURE 6-2. The ideal transformer.

assumed. The basic equation defining the operation of the ideal transformer is

$$\frac{e_p}{e_s} = \frac{N_p}{N_s}, \quad (6\text{-}1)$$

where N_p and N_s are the actual numbers of turns of the primary and secondary windings, respectively.

It is important to note that the polarity of the secondary voltage can be either in phase or 180° out of phase with that of the primary voltage, a relation which is established simply by the manner in which the two coils are wound. In alternating current circuits, this means that the secondary current is a sinusoid, which is either 180° removed (in phase) from the current in the primary or exactly in phase with the primary current. In pulse circuits, the situation is such that the secondary pulse is either the same polarity or the opposite polarity to the input (primary) pulse. It is not necessary to buy a special transformer to reverse the polarity of the output. When one has the actual transformer on hand, he need simply connect the terminals of the desired polarity. For schematic diagrams and illustrations like Fig. 6-2, it is common to indicate the points of like polarity by heavy dots.

Just as the primary and secondary voltages are related in the ideal transformer, so are the primary and secondary currents. The currents, however, are related to the inverse of the turns ratio,

$$\frac{i_p}{i_s} = \frac{N_s}{N_p}. \quad (6\text{-}2)$$

This is a relation that is clear since the transformer is a passive device and since, therefore, the output power can at best equal the input power. Simply stated, power in = power out

$$e_p i_p = e_s i_s \text{ (ideal assumption),}$$

and, by substituting from Eq. (6-1),

$$e_p i_p = \left(\frac{N_s}{N_p} e_p\right) i_s$$

or

$$\frac{i_p}{i_s} = \frac{N_s}{N_p}.$$

Note that Eqs. (6-1) and (6-2) show that secondary circuit actions are related to those of the primary circuit, and vice versa. In other words, a low impedance load in the secondary will result in a higher primary current than a high impedance in the secondary. The following exercise permits us to develop a basic equivalent circuit for the ideal transformer.

Referring to Fig. 6-2, note that we can write

$$Z_L = \frac{e_s}{i_s}. \tag{6-3}$$

And from Eqs. (6-1) and (6-2), we can substitute for e_s and i_s, resulting in

$$Z_L = \frac{N_s/(N_p)e_p}{N_p/(N_s)i_p} = \left(\frac{N_s}{N_p}\right)^2 \frac{e_p}{i_p} \tag{6-4}$$

or

$$Z_L = a^2 Z_p, \tag{6-5}$$

where we have substituted a for the overall turns ratio of the transformer ($a = N_s/N_p$) and Z_p for an impedance that is apparent in the primary side of the transformer as a result of Z_L actually being connected in the secondary. Observe that the turns ratio a is equal to the square root of the impedance ratio:

$$\sqrt{\frac{Z_L}{Z_p}}. \tag{6-5a}$$

The utility of the above transformation is seen from Fig. 6-3, which shows the equivalent circuit for the ideal transformer. This simple circuit shows that, if the assumption of an ideal transformer can be made, then all that is required of the circuit designer in order for him to determine the proper circuit and voltage values for his designs is knowledge of the secondary impedance and the transformer turns ratio.

Example 6-1 Consider the circuit of Fig. 6-2 to determine the following values. The circuit values are: $e_p = 100 \sin 377t$, $Z_L = 1 \text{ k}\Omega$, $a = 4$, and assume the transformer to be ideal.

The Ideal Pulse Transformer

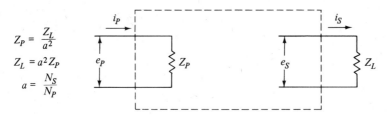

$$Z_P = \frac{Z_L}{a^2}$$
$$Z_L = a^2 Z_P$$
$$a = \frac{N_S}{N_P}$$

FIGURE 6-3. Equivalent circuit for the ideal transformer and related formulas.

Find: (a) e_s
  (b) $i_{s(\text{RMS})}$
  (c) i_p
  (d) Z_p
  (e) P_L (power in load)

Solution: Observe that

$$a = \frac{N_s}{N_p} = \frac{e_s}{e_p},$$

$$4 = \frac{e_s}{100 \sin 377t},$$

(a) $$e_s = 400 \sin 377t.$$

Recall that the equation $E_{\max} \sin \omega t$ defines a sine wave such that $E_{\max}$ is amplitude and ω is angular velocity:

$$\omega = 2\pi F$$

$$F = \frac{\omega}{2\pi} = 60 \text{ Hz}.$$

Thus, the secondary voltage is 400 V maximum at a frequency of 60 Hz.

(b) $$i_{s(\text{RMS})} = \frac{e_{s(\text{RMS})}}{Z_L}.$$

Therefore,

$$i_{s(\text{RMS})} = \frac{0.707(400)}{1 \text{ k}\Omega} = 283 \text{ mA}.$$

(c) $$i_p = \frac{e_p}{Z_p} \quad \text{or} \quad i_p = ai_s,$$

and

$$i_p = 4(283 \text{ mA}) = 1132 \text{ mA} = 1.132 \text{ A}.$$

(d) $$Z_p = \frac{Z_L}{a^2} = \frac{1\ \text{k}\Omega}{16} = 62.5\ \Omega$$

or

$$Z_p = \frac{e_{p(\text{RMS})}}{i_{p(\text{RMS})}} = \frac{(.707)100}{1.132\ \text{A}} = 62.5\ \Omega.$$

(e) $$P_L = \frac{(E_{L(\text{RMS})})^2}{Z_L} = \frac{(283)^2}{1\ \text{k}\Omega} = 80.1\ \text{W}.$$

6-2 THE NONIDEAL PULSE TRANSFORMER

Most often, rough calculations are made, using the ideal assumptions discussed above, and then, later, more practical considerations are made. In the case of the transformer, practical considerations include: compensation for eddy currents and hysteresis effects in the transformer core; leakage impedance, which results from the fact that not all of the flux is confined to the core; the actual wiring resistances associated with the primary and secondary windings; and, finally, the interwiring capacitance. Our concern with these factors in pulse circuits will be to insure that the pulse rise time (high-frequency response) and pulse shape are maintained to some acceptable degree.

All of the important factors might be shown in an equivalent circuit similar to that of Fig. 6-4. Tabulated below, they represent:

R_p — Primary winding resistance
L_p — Primary winding leakage reactance
R_M — Resistive core losses
L_M — Reactive core losses
C — Total primary and secondary winding capacitance
R_s — Secondary winding resistance
L_s — Secondary winding leakage reactance.

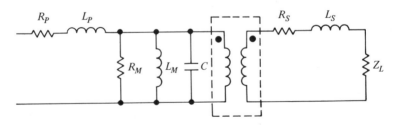

FIGURE 6-4. Equivalent circuit for the nonideal transformer.

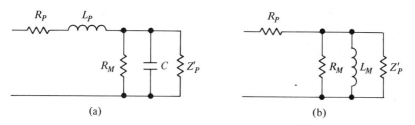

FIGURE 6-5. (a) High-frequency equivalent circuit; (b) low-frequency equivalent circuit.

From the equivalent circuit it is easy to see that some of the factors will affect high-frequency response and others low-frequency response. In pulse circuits, high-frequency response is equated to pulse rise time and low-frequency response to the flat top (tilt) of the pulse.

By referring all impedance to the primary side of the transformer, as we did for the ideal case, and by making some reasonable circuit assumptions, we can construct two separate equivalent circuits for the high- and low-frequency situation, respectively. These equivalent circuits are shown in Fig. 6-5.

In the circuit of Fig. 6-5(a), shunt inductances are neglected, since for high-frequency considerations they represent large impedances and have negligible circuit effects compared to the shunt capacitance. In this circuit, Z_p' represents the total secondary impedance as it is reflected into the primary, or

$$Z_p' = \frac{(R_s + X_{L_s} + Z_L)}{a^2}, \qquad a = \frac{N_s}{N_p}. \qquad (6\text{-}6)$$

The remaining parameters are the same as their counterparts in Fig. 6-4.

The circuit of Fig. 6-5(b) is a low-frequency equivalent circuit. This circuit reflects the fact that shunt capacitance and series inductance become negligible under low-frequency considerations.

In conclusion, it must be remembered that transformers in pulse circuits can be reduced to meaningful equivalent circuits. It is the problem of the circuit designer to insure that reasonable assumptions are made and that the resultant equivalent circuit is realistic for the particular case being considered. Once reduced, the equivalent circuit is attacked with basic circuit analysis techniques such as Kirchhoff's laws, Thévenin's theorem, and so on—which, in pulse circuits, frequently involve higher-level mathematics. In many cases, convenient nomographs and/or charts are available for the solution of such problems.

6-3 THE MONOSTABLE BLOCKING OSCILLATOR

Consider the circuit of Fig. 6-6(a). It is a common configuration for a triggerable blocking oscillator using emitter timing. The transformer employs a tertiary winding to couple the output pulse, but the connection of this winding is of little importance to the actual circuit operation. The remaining two windings of the transformer must be connected, as shown, to provide positive feedback from the collector to the base with a gain (turns ratio) of a.

At rest, the transistor is biased Off by the small negative base bias voltage, V_{BB}. Prior to the time when a trigger is applied (t_0), the state of the circuit is the normal Off condition as seen from the waveforms in Fig. 6-6(b) and from the transistor characteristics in Fig. 6-6(c).

At t_0 the required positive trigger is coupled to the base, and the transistor comes out of cutoff. Collector current increases, and the collector is coupled, via the transformer, to the base as a positive signal (positive feedback) insuring that the transistor is driven into saturation. The trigger is no longer required because of regeneration, and the transistor operating point moves along the saturation part of the characteristics toward point A, reaching that point at time t_A. That is, the transistor collector voltage was high (equal to V_{CC}) and will drop to a low value as collector current (I_C) increases (see arrows on curve). As the transistor enters its active region at point A, collector current starts to drop rapidly, collector voltage rises sharply, and again the regenerative action of the transformer takes over. The flux of the transformer begins to collapse, causing the voltage to reverse across the coil, and the collector voltage rises sharply, driving the transistor rapidly into cutoff. This completes the generation of the output pulse.

The inductive effects of the transformer cause an overshoot, which could possibly be large enough to damage the transistor. In circuit design, careful consideration must be given to this overshoot to insure that neither the base emitter breakdown voltage nor the collector emitter breakdown voltage is excluded. Frequently, a diode is connected in the collector circuit, in shunt with the transformer, to clip the overshoot.

Another effect that must be considered is the improper damping of the overshoot. If the overshoot is permitted to go into cyclic oscillation, the first positive-going swing in the base circuit could cause a retriggering and, essentially, result in a free-running circuit. It is common in circuit design to place an external resistor across the transformer for additional damping to insure monostable operation.

The Monostable Blocking Oscillator

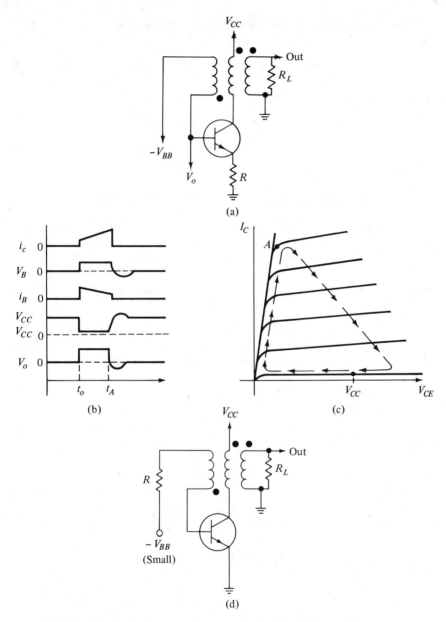

FIGURE 6-6. (a) Emitter timed monostable blocking oscillator; (b) the circuit waveforms; (c) movement of the operating point during the generation of a pulse; (d) base timed monostable blocking oscillator.

It can be shown that the duration of the output pulse for the circuit above is

$$t \cong \frac{aL}{R} - \frac{a_1^2 L}{R_L}, \qquad (6\text{-}7)$$

where R_L = load resistance,

a_1 = tertiary-to-collector turns ratio $\left(\dfrac{N_{\text{ter}}}{N_C}\right)$,

a = base-to-collector turns ratio $\left(\dfrac{N_B}{N_C}\right)$,

L = magnetizing inductance of collector winding.

Equation (6-7) points out that, for the particular circuit configuration shown in Fig. 6-6, the pulse output is independent of any transistor characteristics, notably, h_{FE}. This is an important point, because of the sensitivity of transistor parameters to temperature variation. For this reason the "emitter timed" blocking oscillator has a definite advantage over the "base timed" blocking oscillator of Fig. 6-6(d).

Example 6-2 Find the stated quantities for the pulse transformer of Fig. 6-6(a) if the circuit parameters are 12 V, 100 Ω, 300 Ω, 100 mH, 100 μsec, and 2 for V_{CC}, R, R_L, L, t, and a_1, respectively. Also, $N_C = 75$ turns.
Find: (a) a
(b) N_B
(c) N_{ter}

Solution: From Eq. (6-7),

$$t = \frac{aL}{R} - \frac{a_1^2 L}{R_L}.$$

Therefore,

$$(100)10^{-6} = \frac{a(.1)}{100} - \frac{(4)(.1)}{300}$$

$$a = 1.433.$$

Also,

$$a = \frac{N_B}{N_C}, \quad N_B = (1.433)(75) = 107 \text{ turns}$$

and

$$a_1 = \frac{N_{\text{ter}}}{N_C}, \quad N_{\text{ter}} = 150 \text{ turns}.$$

6-4 TRIGGERING THE BLOCKING OSCILLATOR

The blocking oscillator is easily triggered, the only requirement being that the transistor must be brought out of cutoff. Where pulses are used, it is desirable to insure that the trailing edge of the trigger pulse does not generate a spike that will drive the transistor back to cutoff before the desired output pulse width is obtained from the blocking oscillator. One way to insure this is to make the trigger pulse width greater than the desired output pulse width. Also, through the use of a simple RC integrator, the input trigger pulse shape can be altered to the point where a sharp trailing edge spike does

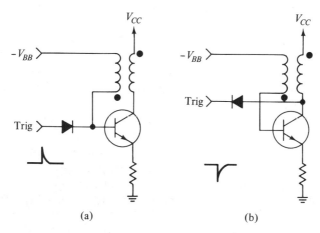

FIGURE 6-7. Methods for triggering the blocking oscillator: (a) positive triggers to the base; (b) negative triggers to the collector.

not exist. Finally, diode clippers can be employed or even an active transistor stage can be used to couple the trigger and prevent feedback from the blocking oscillator to the trigger source. Two other methods of triggering the monostable circuit are shown in Fig. 6-7.

6-5 THE FREE-RUNNING (ASTABLE) BLOCKING OSCILLATOR

The free-running, or astable, configuration of the blocking oscillator circuit is very similar to the monostable circuit just discussed. We have already noted that were it not for proper clamping of the overshoot (see Fig. 6-6), the monostable circuit could easily retrigger itself. The astable circuit therefore is one that insures continual retriggering, usually by the insertion of a capacitor to control timing. Also, the emitter base bias source is reversed to insure that the transistor will not stay cut off, although cutoff was desired in the monostable case. Observe the circuit of Fig. 6-8(a). To emphasize that many variations of the blocking oscillator are to be found, the regenerative loop couples the emitter and collector circuits, whereas the collector and base circuits were coupled in the preceding case. This will not alter the basic action of the blocking oscillator during the pulse interval that was described for the circuit of Fig. 6-6. Note, however, that because feedback is from collector to emitter, the proper polarity of the transformer connections are such as to provide an in-phase (same polarity) pulse to the emitter, where an opposite polarity pulse was previously fed to the base.

During the generation of the pulse, operation is identical to that for the monostable case. However, in the circuit of Fig. 6-8, while the pulse is being generated, the capacitor C charges positively on the side connected to the emitter, as a result of the large emitter current during the pulse. When the regenerate action causes I_C to saturate, the pulse transformer flux starts to collapse, and the voltage across the base winding reverses. This voltage and the residual change on C cuts off the transistor. The charge on the capacitor will gradually leak off, depending on the RC time constant. As the emitter (capacitor voltage) potential approaches $V_{BE(\text{on})}$ volts, the transistor comes out of cutoff, and the next pulse is generated. The potential $-V_{EE}$ insures that the transistor will come out of cutoff and maintains the On condition until the negative pulse from the collector is coupled to the emitter to drive the transistor into saturation during the generation of the pulse. The critical circuit waveforms are shown in Fig. 6-8(b). E_{CM} represents the maximum voltage that charges the capacitor during transistor conduction.

For most applications the actual pulse is small compared to the interval between pulses. A good approximation to the overall cycle time is the time between the trailing edge of one pulse and the leading edge of the next—or in other words, the cutoff time that is controlled by the RC combination. This time is given by

$$t = RC \ln \frac{V_{EE} + aV_{CC}}{V_{EE}} \quad (V_{EE} \text{ and } V_{CC} \text{ are absolute values}). \quad (6\text{-}8)$$

The quantity aV_{CC} is a close approximation for the exact value. The determination of this value requires the assumption of a constant charging current for the capacitor and knowledge of the pulse width of the single-form transformer. The input parameters of the transistor would also be needed. As with all blocking oscillator circuits, there are many assumptions required to do a complete analysis, and the results, as previously indicated, are approximate.

The Free-Running (Astable) Blocking Oscillator

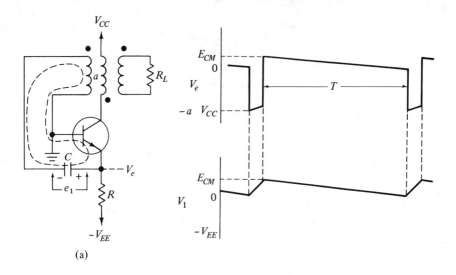

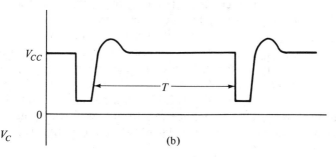

FIGURE 6-8. (a) An emitter coupled astable blocking oscillator; (b) capacitor and emitter waveforms for the circuit.

Example 6-3 Determine the approximate frequency for the oscillator in Fig. 6-8(a). The circuit values are: $R = 3.3 \text{ k}\Omega$, $V_{EE} = -4$ V, $a = 2$, $V_{CC} = +8$ V, $C = 1000$ pF.

Solution: Eq. (6-8) is

$$t = RC \ln \frac{V_{EE} + aV_{CC}}{V_{EE}}$$

$$t = (3.3)(10^3)(10^3)(10^{-12}) \ln \frac{4 + 16}{4}$$

$$= 5.31 \times 10^{-6} \text{ sec}.$$

Therefore,

$$F = \frac{1}{t}$$

$$= \frac{10^6}{5.3} = 0.188 \times 10^6$$

$$= 188 \text{ kHz}.$$

6-6 SYNCHRONIZATION OF THE ASTABLE BLOCKING OSCILLATOR

The circuit of Fig. 6-8 can be used as a dividing circuit or a highly stable pulse repetition frequency generator by employing synchronization. Note from the waveforms of Fig. 6-8(b) that the repetition frequency is determined by the capacitor waveform. When this voltage reaches zero a new pulse is generated. The shallow slope of this voltage, coupled with the fact that transistor and circuit parameters are somewhat sensitive to temperature variations, indicates that the repetition frequency of a circuit like that in Fig. 6-8 can be unstable. One possible technique for correcting this shortcoming is to use a stable synchronized trigger to trigger the oscillator just before it reaches the end of its natural cycle. The fact that this trigger can be derived from the blocking oscillator itself distinguishes the circuit from the simple triggered monostable circuit.

Note the block diagram of Fig. 6-9. If the output pulse from the blocking oscillator itself is fed to a delay line—and from there back to the emitter of the blocking oscillator (see Fig. 6-8), then, before the emitter voltage decays

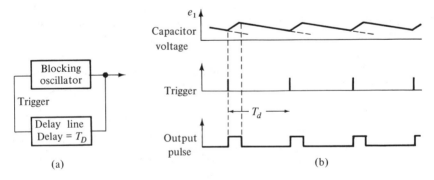

FIGURE 6-9. (a) A delay-line stabilized blocking oscillator; (b) circuit waveforms.

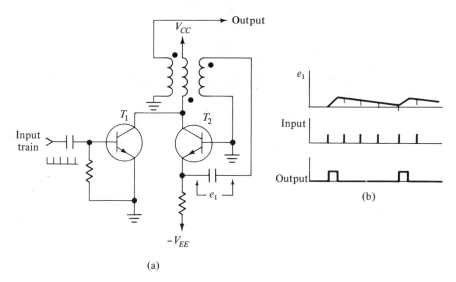

FIGURE 6-10. (a) Higher frequency pulse train applied to a blocking oscillator through a separate transistor stage; (b) circuit waveforms showing output pulses at one-quarter the rate of input pulses.

to zero, the delayed trigger causes the blocking oscillator to recycle. In this manner, the highly stable delay line determines the pulse repetition period, T_d. The only requirement for this scheme is that the delay line period be less than the natural period of the blocking oscillator, and that the trigger be sufficient in amplitude to overcome the cutoff bias and drive the transistor into saturation. The delay trigger, as shown in Fig. 6-9(b), has a different shape than the output pulse. This may be desirable and can be accomplished easily with a simple differentiating circuit.

If, in place of the delay trigger in Fig. 6-9(b), a pulse train with a higher repetition frequency were to exist, then it could be possible to use the blocking oscillator to count down, or divide, this higher frequency. The blocking oscillator triggers only when the combination of trigger pulse amplitude and capacitor voltage is sufficient to overcome the cutoff bias. This feature is illustrated in Fig. 6-10, which shows that, for this particular circuit arrangement, an output pulse is generated for every fourth input, or in other words, we have a divider-by-four circuit.

Review Questions

1. State as many factors as possible that affect the frequency of an astable blocking oscillator.
2. Why is a blocking oscillator sometimes referred to as a relaxation oscillator?
3. Explain how a positive input pulse will trigger the circuit of Fig. 6-7(a).
4. Define an "ideal" pulse transformer.
5. How do Z_p, Z_L, and turns ratio a relate to each other? Explain.
6. Explain the meaning of the circuit parameters R_p, R_m, C, and R_s as illustrated in Fig. 6-4.
7. Discuss the disadvantages of the circuit in Fig. 6-6(d) as compared to the circuit of Fig. 6-6(a).
8. Explain the meaning of the heavy dots on a transformer schematic diagram.
9. Explain the circuit operation when triggering a PNP astable blocking oscillator.
10. What change in the frequency of the circuit illustrated by Fig. 6-8 would occur if V_{EE} were increased? Explain.
11. Under what circuit condition, if any, would a positive pulse trigger the oscillator of Fig. 6-7(b)?
12. What is the meaning of the term *tertiary winding*?
13. Define the term *turns ratio*.
14. What is the meaning of the arrows on the graph of Fig. 6-6(c)? Explain in detail.
15. Explain the action of the delay line of Fig. 6-9(a).

Problems

1. Draw a schematic diagram of a PNP base timed astable blocking oscillator.
2. The transformer circuit of Fig. 6-2 has the following values: $Z_L = 50 \, \Omega$, $a = .2$, and $P_L = 12.5$ W.
   Find: (a) Z_p
   (b) $e_{p(max)}$
   (c) $e_{p(RMS)}$
   (d) I_p
   (e) I_s

Problems

3. Refer to the circuit of Fig. 6-6(a) to solve the problem. The circuit values are: $R_L = 50\,\Omega$, $N_{ter} = 50$, $N_C = 100$, $N_B = 200$, and $t/L = .006$.
   Find: (a) R
   (b) a_1
   (c) a
4. Solve for the pulse duration t of Problem 3 if $L = 100$ mH and $R = 300\,\Omega$. All other values remain the same.
5. Draw a schematic diagram of a PNP emitter timed monostable blocking oscillator. Show input trigger pulses.
6. Explain the operation of transistor T_1 in the circuit of Fig. 6-10(a).
7. Determine the output pulse frequency of Fig. 6-8(a) if the circuit values are: $R_L = 100\,\Omega$, $V_{CC} = 12$ V, $V_{EE} = -6$ V, $C = .001\,\mu$F, $R = 500\,\Omega$, and $a = \frac{1}{2}$.
8. Solve for the value of a that is needed to change the frequency of the blocking oscillator in Example 6-3 to 150 kHz. All other values remain the same.
9. Determine the pulse duration t for the circuit of Fig. 6-6(a) if the circuit parameters are: $V_{CC} = 14$ V, $R_L = 100\,\Omega$, $a_1 = \frac{1}{3}$, $a = 2$, $L = 50$ mH, $R = 1$ kΩ.
10. Show by a schematic diagram how to trigger the circuit of Fig. 6-6(a) using a diode input. Illustrate the polarity and necessary amplitude of the input pulse.

Time Base Generators

7

A time base generator, frequently called a sweep circuit, is a circuit that has an output that is a predictable function of time. Most frequently it is desirable that this function of time be linear; however, we shall see that some degree of nonlinearity is tolerable in most cases. The ideal output waveform is commonly called a "sawtooth" or "ramp" function and is shown as in Fig. 7-1(a). Figure 7-1(b) shows a typical waveform in which the total period T actually consists of the sweep portion (τ_1) and the flyback or retrace (τ_2).

Sweep circuits have many useful applications in electronics. You are probably familiar with the sweep generator in an oscilloscope that moves the electron beam across the face of the scope in a linear manner. Similar circuits are used in TV sets and radar systems. Frequently a sweep circuit with variable period T is used as a delay circuit for a trigger. In this application, the primary trigger initiates the sweep signal and the trailing edge of the sawtooth

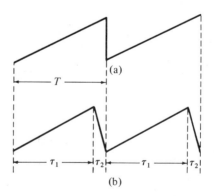

FIGURE 7-1. Sawtooth waveform: (a) ideal; (b) actual.

Sweep Circuit Linearity

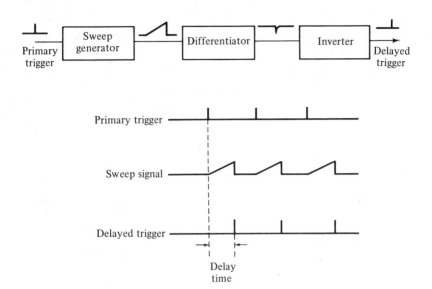

FIGURE 7-2. A block diagram of a delay trigger generator and waveforms.

causes the delayed trigger. A block diagram of this scheme is shown in Fig. 7-2.

7-1 SWEEP CIRCUIT LINEARITY

The assumption is that a linear voltage-vs-time relation is desired in sweep circuits. The following remarks hold true for this case.

A perfectly linear waveform is one in which the rate of change of voltage with respect to time (slope) is constant for the entire duration of the usable part of the signal. This discussion is not concerned with linearity during the nonusable part of the signal, which is commonly called the retrace or flyback (t_2 of Fig. 7-1b).

The equation for a perfectly linear waveform would be the equation of a straight line with a constant slope. Recall from algebra that a straight line is formed by the equation $y = mx + b$ or, in our case,

$$e = kt + c. \tag{7-1}$$

An example of a circuit with a nonlinear voltage-vs-time equation is the simple capacitor changing characteristics of Eq. (1-9):

$$e_C = E(1 - e^{-t/RC}).$$

The deviation of the capacitor change curves from the straight line of Eq. (7-1) can be specified in several ways. One such measurement of nonlinearity is called slope error and is defined as

$$\text{Percent slope error} = \frac{\text{Initial slope} - \text{Final slope}}{\text{Initial slope}} \times 100. \quad (7\text{-}2)$$

Another measurement of nonlinearity is called displacement error and is defined as

$$\text{Percent displacement error} = \frac{\text{Maximum voltage difference}}{\text{Maximum voltage}} \times 100. \quad (7\text{-}3)$$

The following example illustrates the difference between slope error and displacement error.

Example 7-1 Calculate the slope error and displacement error for the capacitor charge curve resulting from the circuit in Fig. 7-3. Assume the output voltage is to be used for only one time constant.

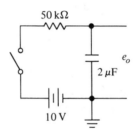

FIGURE 7-3. *RC* network.

Solution: (1) One time constant $= RC = (50 \text{ k}\Omega)(2 \mu\text{F}) = 100$ msec. (2) In one time constant the capacitor will change to 63.4 percent of the final voltage, or 6.34 V. (3) The equation for voltage is $e_o = E(1 - e^{-t/RC})$. The slope of the charging curve at any time t is

$$(\text{Rate of voltage change with respect to time}) = \text{slope} = \frac{E}{RC} e^{-t/RC}. \quad (7\text{-}3a)$$

The initial slope occurs at time $t = 0$,

$$\text{Initial slope} = \frac{E}{RC} e^{-0} = \frac{E}{RC}.$$

The final slope occurs at $t = RC$,

Sweep Circuit Linearity

$$\text{Final slope} = \frac{E}{RC} e^{-1} = \frac{E}{RCe}.$$

The percent slope error is

$$\text{Percent slope error} = \frac{E/RC - E/RCe}{E/RC} \times 100$$

$$= \left(1 - \frac{1}{e}\right)(100)$$

$$= 63.4 \text{ percent.}$$

The displacement error can be found most easily by using the universal time constant curve (Fig. 1-4). This is accomplished by tracing the curve on a sheet of graph paper and plotting the straight line as shown in Fig. 7-4. Note that at $t/RC = .45$ (approximately) the difference between the two curves is maximum and is about .8 V.

$$\text{The displacement error} = \frac{\text{Maximum difference}}{\text{Maximum voltage}} \times 100 = \frac{.8}{6.34}(100)$$

$$= 12.6 \text{ percent.}$$

Although RC circuits do not exhibit a great deal of linearity, they are frequently used in simple timing operations. Linearity is kept within very reasonable limits by employing only a small portion of the charging curve. (See Section 1-7, Chapter 1.)

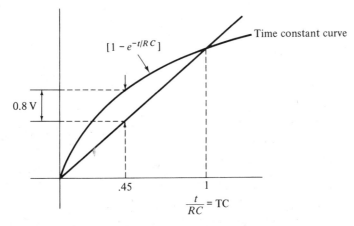

FIGURE 7-4. Displacement error.

7-2 THE GAS TUBE SWEEP CIRCUIT

A basic extension of the simple RC circuit provides for an automatic switch that terminates the charging of the capacitor and at the same time provides a discharge path to complete the sawtooth period. A convenient device that can act as the switch is a gas tube or common neon bulb.

A typical current-voltage curve for a common neon bulb is shown in Fig. 7-5. Until the voltage is increased to the ignition potential E_I, no current flows through the tube. When the tube fires, the voltage drops back to the quiescent potential E and the tube acts like a short circuit. As the potential across the fired tube is decreased, the current traces out the dotted portion of the curve in Fig. 7-5 until the extinction potential (E_D) is reached and the tube becomes an open circuit.

When employed in a circuit like that of Fig. 7-6, the neon bulb becomes the switching element for the most basic of all sawtooth generators. In this circuit, the proper choice of E_{BB} and R will insure a free-running oscillation, while the choice of C will determine the frequency.

The basic circuit operation is relatively easy to understand. Current from the battery flows through R and charges the capacitor exponentially with a time constant of RC. When the capacitor potential reaches the ionization potential of the neon bulb, it fires and conducts heavily, effectively placing a short circuit across the capacitor. The capacitor discharges through the

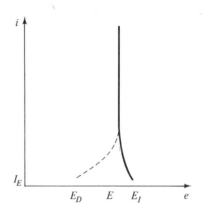

FIGURE 7-5. Voltage-current characteristic for neon bulb.

The Gas Tube Sweep Circuit

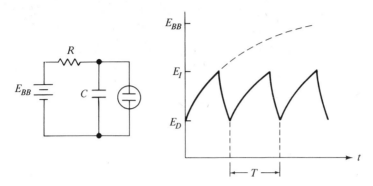

FIGURE 7-6. The basic sawtooth generator.

short circuit until the potential drops to the de-ionization potential of the neon bulb, becoming an open circuit. The time constant during this part of the cycle is very short because of the near zero resistance of the conducting bulb.

It is obvious from the description of circuit operation that E_{BB} must be greater than E_I. However, if the bulb is to de-ionize, the battery cannot be too large or the resistor too small. In other words, the extinction potential must be greater than the difference between the supply voltage and the drop across the resistor,

$$E_D > E_{BB} - IR_D, \qquad (7\text{-}4)$$

or the bulb will not extinguish.

The total period of the sawtooth consists of the charge and discharge time of the capacitor—however, the discharge time is small and can be neglected. Under this assumption, then, the period would be the difference between the time required to charge to E_D and the time to charge to E_I, using RC as the effective time constant. In other words, the period of concern is the time difference between two points on the charging curve for the capacitor, or

$$T \cong RC \ln\left(\frac{E_{BB} - E_D}{E_{BB} - E_I}\right). \qquad (7\text{-}5)$$

Example 7-2 Determine the frequency of the gas tube oscillator of Fig. 7-6. The circuit values are: $R = 100$ kΩ, $C = .01$ μF, $E_I = 67$ V, $E_D = 52$ V, and $E_{BB} = 200$ V.

Solution: The period of oscillation is given in Eq. (7-5); thus,

$$T = RC \ln\left[\frac{E_{BB} - E_D}{E_{BB} - E_I}\right]$$

$$T = (10^5)(10^{-8}) \ln\left[\frac{200 - 52}{100 - 67}\right]$$

$$T = 113 \ \mu\text{sec}$$

$$F = \frac{1}{T} = 8.8 \ \text{kHz}.$$

7-3 THE THYRATRON SWEEP CIRCUIT

A basic disadvantage of the neon bulb circuit is lack of control, for once power is applied, the circuit operates continuously until the power is removed. Additionally, the voltage swing is quite small and unalterable. There are other disadvantages to the basic gas tube circuit, but the ones mentioned above can be eliminated by replacing the neon bulb with a gas triode or thyratron.

The plate characteristic for a thyratron is quite similar to the curve for the gas tube in that the tube conducts readily after the ionization potential is reached and in that the conduction potential is quite constant. The ionization potential of the thyratron is controllable, however, and this feature makes it possible to overcome the disadvantages of the neon bulb circuit. The control element in the thyratron is the grid. A typical graph of grid-control voltage vs ionization potential is shown in Fig. 7-7. We see from this graph that, as the grid is made more negative, greater plate potentials are required to "fire" the tube. Once the tube fires, however, the plate potential drops to a low constant value until the current falls below the minimum value required to sustain ionization, I_D.

An adjustable thyratron sweep circuit is shown in Fig. 7-8. In this circuit, the setting of the control potentiometer R_C determines the grid-control voltage. Once a setting is selected, the circuit operation is identical to that of the neon bulb circuit, except that the additional resistance R_2 in the plate circuit must be included in the calculations. Resistor R_1 must be selected to provide a rapid discharge for C, and yet its value must not be so small that it permits an excessively high discharge current through the thyratron. C and R_2 form the charge time constant for the circuit and $R_1 \ll R_2$. The required condition for oscillation is

$$E_I < E_{BB} < (E_D + I_D R_2 + I_D R_1), \tag{7-6}$$

and the period of oscillation (assuming negligible flyback time) is

$$T = R_2 C \ln\left(\frac{E_{BB} - E_D - I_D R_1}{E_{BB} - E_I}\right). \tag{7-7}$$

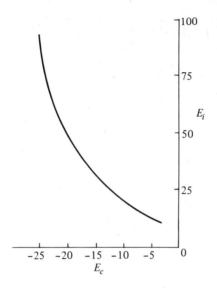

FIGURE 7-7. Thyratron control voltage vs. ionization potential.

The peak-to-peak amplitude of the output waveform is

$$e_{p\text{-}p} = E_I - E_D - I_D R_1. \tag{7-8}$$

Most thyratrons require about 1 mA of current to sustain ionization, and R_1 is calculated on the basis of the maximum discharge current; therefore, the product $I_D R_1$ is usually small and may be neglected in Eqs. (7-6) and (7-7).

Example 7-3 Determine the following quantities of the thyratron oscillator in Fig. 7-8. The circuit values are 500 Ω, 50 kΩ, .05 μF, 200 V, 900 μA, 20 V, and 23 V for R_1, R_2, C, E_{BB}, I_D, E_D, and E_C, respectively. Assume that the thyratron characteristic curve is that of Fig. 7-7.

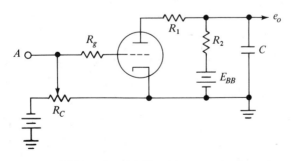

FIGURE 7-8. Adjustable sweep circuit.

Find: (a) E_I
(b) I (peak discharge)
(c) frequency

Solution: E_I is found from the graph of Fig. 7-7,

$$E_I = 75 \text{ V}.$$

Thus,

$$t = R_2 C \ln \left[\frac{E_{BB} - E_D - I_D R_1}{E_{BB} - E_I} \right]$$

$$= (2.50)(10^{-3}) \ln \left[\frac{200 - 19.55}{200 - 75} \right]$$

$$= 910 \text{ } \mu\text{sec}.$$

Therefore,

$$F = \frac{1}{T}$$

$$= \frac{10^6}{910} = 1.09 \text{ kHz},$$

and

$$I \text{ (peak discharge)} = \frac{E_I - E_D}{R_1} = \frac{(75 - 20) \text{ V}}{500 \text{ }\Omega} = 110 \text{ mA}.$$

7-4 SYNCHRONIZATION OF THE THYRATRON SWEEP GENERATOR

If a stable sinusoid were to be fed into the grid circuit of Fig. 7-8 at point A, the effect would be to improve the stability of the output signal by synchronizing it with the input sinusoid. Figure 7-9 illustrates the operation.

Capacitor C charges toward E_{BB} just as in the gas tube circuit. Without the synchronized signal present, the tube will fire when C charges to E_I (a

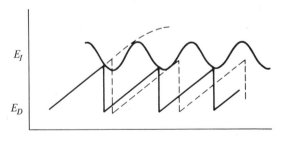

FIGURE 7-9. Synchronization of the thyratron circuit.

The Triode Integrator

value determined by the control voltage). The waveform will be that shown by the dotted lines. Now, when the sine wave is added to the control voltage, it will cause a periodic fluctuation (E_D) in the ionization potential (E_I) (as shown in Fig. 7-9), and the sweep voltage will lock to the sinusoidal rate. The value of such a scheme is twofold. First, the sweep can be synchronized to a stable source (which helps to eliminate sweep period fluctuations resulting from minor circuit changes), and second, the phase of the sweep signal becomes adjustable within a certain range of amplitude of the synchronized signal.

7-5 THE TRIODE INTEGRATOR

The triode integrator is simply an active circuit that takes a square wave input signal and provides an output that is an integrated square wave, that is, a ramp function. Because the output is a ramp function, the circuit can be placed in the category of time base generators. See Fig. 7-10. With no signal present, the grid maintains the tube in saturation because of the coupling to $B+$ via R_g. A negative-step input switches the tube to cutoff—the plate voltage tries to rise, but the capacitor, in parallel with the tube, prevents a rapid rise in voltage. Instead, current is drawn through R to charge C, producing the output waveform shown in Fig. 7-9. Keeping RC long with respect to T provides for reasonable linearity, similar to that for the RC circuit discussed earlier. Removal of the input step drives the tube back into conduction. The capacitor discharges through the parallel combination of R and the dynamic plate resistance of the tube, r_p. Since r_p is much less than

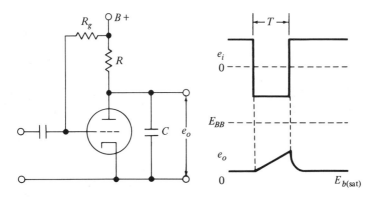

FIGURE 7-10. The triode integrator with waveforms.

R, the discharge time constant is much less than the charge time constant, and an effective sawtooth is generated.

7-6 THE TRANSISTOR INTEGRATOR

The transistor integrator is nearly identical to its triode counterpart, with two exceptions. First, the flyback part of the waveform is affected because there is not complete isolation of the output from the input as there is in the triode circuit, and second, with the transistor we have the option of choosing a PNP or an NPN device.

The circuit is shown in Fig. 7-11, where the active device in the triangle (A) could be either type of transistor. For our purpose we will assume a PNP transistor, which will require a positive-gating pulse and will result in a negative-going output.

The circuit operation is as follows. When the gating pulse brings the transistor out of saturation at t_0, collector current drops to nearly zero, and the capacitor charges from the saturation value of collector voltage toward the supply potential $-V_{CC}$ through R_C. If the gate duration T is short relative to the time constant $R_C C$, then a nearly linear ramp is obtained as an output v_o.

$$v_o = -V_{CC}(1 - e^{-t/RC}).$$

When the gate pulse is removed at t_1, the emitter base junction is forward biased and base current flows through R_b to V_{CC}. This base current is amplified and appears in the collector circuit as $h_{FE}I_B$. Also, C discharges through

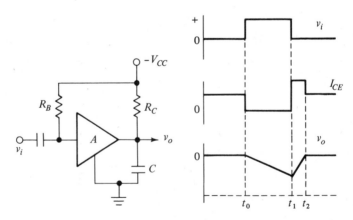

FIGURE 7-11. The transistor integrator with waveforms.

The Miller Integrator

the saturated transistor and provides a current greater than the original saturation current. During the flyback portion, the equation that governs the circuit operations is

$$v_o = -V_{CC} + h_{fe}I_B R_C - h_{fe}I_B R_C e^{-t/RC}, \qquad (7\text{-}9)$$

where t is the time measured from the end of the gating pulse.

7-7 THE MILLER INTEGRATOR

The Miller integrator shown in Fig. 7-12 can best be understood by reviewing the theory of Miller effect in an amplifier. Figure 7-12(a) represents a simplified ac equivalent circuit of a PNP common emitter amplifier. Looking into the input terminals of the amplifier, the answer to the question, "What is the total input capacity of the amplifier?" must be clearly understood. Observe that the total ac current is

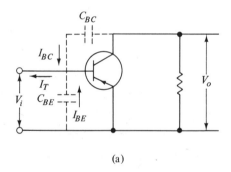

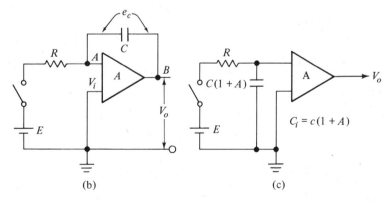

FIGURE 7-12. (a) Circuit to analyze Miller effect; (b) the Miller integrator basic circuit; (c) an equivalent diagram.

and
$$I_T = I_{BC} + I_{BE}$$

$$I_T = \frac{V_{in}}{X_{CT}}.$$

X_{CT} is the total input capacity reactance. Also,

$$I_{BE} = \frac{V_{in}}{X_{C_{BE}}}.$$

Notice that

$$I_{BC} = \frac{V_{BC}}{X_{C_{BC}}} = \frac{V_{in} - V_{out}}{X_{C_{BC}}}.$$

Now V_{out} is V_{in} times the voltage gain of the amplifier. Recall that the common emitter amplifier is an inverting amplifier; therefore, V_{out} is $(-A)(V_{in})$. Substituting this quantity into the I_{BC} equation yields

$$I_{BC} = \frac{V_{in} - (-V_{in}A)}{X_{C_{BC}}}$$

$$I_{BC} = \frac{V_{in}(1 + A)}{X_{C_{BC}}}$$

and

$$\frac{V_{in}}{X_{CT}} = \frac{V_{in}}{X_{C_{BE}}} + \frac{V_{in}(1 + A)}{X_{C_{BC}}}.$$

Dividing out common factors and solving for C_t from this equation gives

$$C_t = C_{BE} + C_{BC}(1 + A). \tag{7-10}$$

The term $C_{BC}(1 + A)$ is commonly called the Miller capacity, and this apparent capacitance multiplication due to the amplifier voltage gain is referred to as the Miller effect. If an external capacitor is added to C_{CB}, as in Fig. 7-12(b), then the input capacitance becomes quite large, depending upon the value chosen for C. If $C \gg C_{BE}$ or C_{BC}, then Eq. (7-10) may be reduced to equal

$$C_{in} \approx C(1 + A). \tag{7-10a}$$

Therefore, the effect of the capacitor across the amplifier is to charge C_{in}. A new equivalent circuit is shown in Fig. 7-12(c). Now,

$$A = \frac{V_o}{V_i} = \frac{V_o}{V_{C_{in}}} = \frac{V_o}{E(1 - e^{-t/RC_{in}})} \tag{7-11}$$

or

$$V_o = AE(1 - e^{-t/RC_{in}}), \tag{7-12}$$

but since $RC_{in} = RC(1 + A)$, we have an equation for a simple RC circuit with a very long time constant $\tau = RC_{in} = RC(1 + A)$. Since E represents

The Miller Integrator

a constant input signal, the rate of change of the output voltage with respect to time is

$$\frac{dV_o}{dt} = \frac{d}{dt} AE(1 - e^{-t/RC_{in}})$$

$$= \frac{AE}{RC(1+A)} e^{-t/[RC(1+A)]}. \qquad (7\text{-}13)$$

For values of A approximately ∞,

$$\frac{A}{1+A} \cong 1 \quad \text{and} \quad e^{-t/\infty} = 1$$

and for t—small compared to $RC(1 + A)$, a design figure easy to achieve—

$$(\text{slope of output voltage}) = \frac{E}{RC}.$$

The linear equation for the output voltage is

$$V_o = \frac{E}{RC} t. \qquad (7\text{-}14)$$

As seen from Eq. (7-14), V_o is a function of the constants E and RC; therefore, this circuit provides a very linear ramp voltage out, and it has been widely used in many practical sweep circuits.

The Miller integrator requires a high input impedance amplifier. Until recently this could best be achieved with vacuum tubes. However, the field effect transistor also has a high input impedance and behaves very much like the triode. Integrated circuit amplifiers with high gain and high input impedance are built, using several stages with an emitter follower as the input stage. In these cases, the amplifier represented in the triangle of Fig. 7-12 is the integrated circuit Fig. 7-13; the capacitor shunts all of the stages. Overall amplifier gains in excess of several thousand are not uncommon in configurations such as this.

Example 7-4 Determine the stated quantities for the circuit of Fig. 7-12 if the circuit parameters are: $C_{in} = 1\mu F$, $C_1 = 1000$ pF, $R = 100$ kΩ, and E is a 20 V gate pulse of 10 μsec duration. (See Fig. 7-17.)

Find: (a) A
      (b) V_o at the end of gate duration
      (c) τ

Solution: From Eq. (7-10a),

(a) $$C_t = C(1 + A)$$

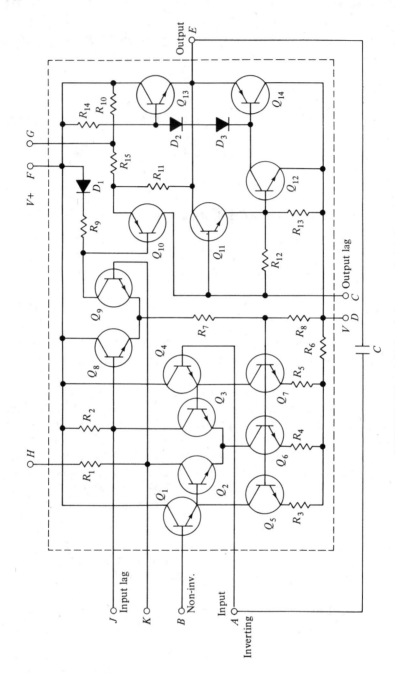

FIGURE 7-13. A high gain integrated circuit.

$$A = \frac{C_t}{C} - 1 = \frac{10^{-6}}{10^{-9}} - 1 = 10^3 - 1$$

$$= 999 \approx 1000.$$

(b) V_o is given by Eq. (7-14):

$$V_o = \frac{E}{RC}t, \quad t = \frac{20 \text{ V}}{10^5 \times 10^{-9}} 10^{-5}$$

$$= 20(10^4)10^{-5} = 20(.10) = 2 \text{ V}.$$

(c) By definition,

$$\tau = RC_{in} = RC(1 + A)$$
$$= 10^5 \times 10^{-6} = .1 \text{ sec}.$$

7-8 MILLER SWEEP CIRCUIT WITH A PENTODE

Frequently a pentode is the amplifier used in a Miller circuit; the constant input voltage is applied in the form of a positive gate to the suppressor. The circuit is shown in Fig. 7-14 along with the significant waveforms at the tube elements.

In the quiescent state, prior to t_0, the control grid potential is near zero as a result of the grid cathode clamp. The tube is in cutoff because of the

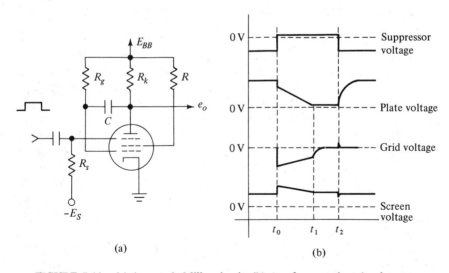

FIGURE 7-14. (a) A pentode Miller circuit; (b) waveforms at the tube elements.

negative potential at the suppressor, and the screen potential is relatively low because the screen draws current from the space charge.

The application of a positive gate to the suppressor causes the tube to conduct. There is an instantaneous drop in plate potential. This drop in plate voltage is transmitted directly to the grid via the capacitor, and from this point, which is above cutoff, the grid potential starts its rise. Because r_p is large, only a small change in grid voltage will be required to cause the plate voltage to bottom out (because the maximum plate current, $i_p \approx E_{RB}/r_p$, is very small), but until this happens, a very linear rundown of plate voltage exists.

When bottoming does occur, the grid rapidly clamps to the zero cathode potential, with the RC time constant being the controlling factor. At this point, the tube is conducting and will continue to do so until the gate is removed from the suppressor. Therefore, from the time that rundown is completed until the gate is removed, the plate (and the capacitor) is clamped to a low potential. Since the plate draws some current during this period, the screen potential is higher than its quiescent value because there is less space current available to the screen.

Removal of the gate at the suppressor cuts the plate current off. The capacitor charges to E_{BB} through r_p. The sharp cutoff of plate current when the gate is removed results in the small spike at the plate, grid, and screen.

7-9 THE PHANTASTRON CIRCUIT

The Phantastron is a frequently used circuit that is a very simple extension of the basic Miller pentode circuit described above. In the Phantastron, the positive gate required at the suppressor is derived from the screen voltage, which, as we note from Fig. 7-14, is an internally generated positive gate. By employing this internal gate, the Phantastron requires only a trigger (instead of a gate) to initiate the Miller rundown.

The basic circuit is shown in Fig. 7-15. In this circuit the negative supply, in conjunction with resistors R_1, R_2, and R_3, maintains the proper quiescent signal levels. The application of a positive trigger to the suppressor initiates the actions by diverting current from the screen to the plate. From this point on, the circuit operation is the normal Miller operation described above, except that the cycle ends when the tube bottoms out, because there is no external gate on the suppressor to maintain plate current.

There are many variations of the Phantastron circuit; however, the basic principles of operation are the same for all. The important features of the

The Bootstrap Sweep Circuit

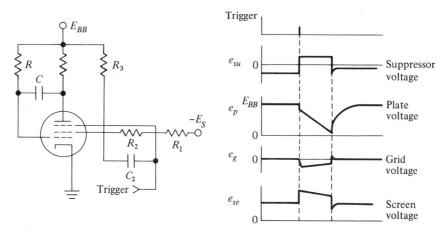

FIGURE 7-15. The phantastron circuit and associated waveforms.

circuit are its linearity of sweep and the fact that sweep time is controllable by varying any of E_{BB}, R, or C. A commonly used arrangement is one where the plate voltage is adjustable by use of a plate catching diode between a variable resistor and the plate of the pentode.

7-10 THE BOOTSTRAP SWEEP CIRCUIT

As previously mentioned, the major shortcoming of all of the sweep circuits that employ capacitor changing is the deviation from linearity that results due to the exponentially decreasing charging current. If the supply voltage to an RC circuit could be increased as the capacitor charge increases, then it would be possible to achieve a linear charging current and hence a linear voltage-time signal on the capacitor. The bootstrap sweep circuit is based on this principle.

The fundamental idea of the bootstrap sweep circuit is illustrated in Fig. 7-16. With switch S closed, there is no input signal to the amplifier and, as a result, no signal out. When the switch is opened, the capacitor will begin to charge, drawing current through R with an initial value of E/R. However, the capacitor voltage e_C is also the input signal to the amplifier. The result is, that as capacitor voltage is developed, it is added to the supply voltage. Thus, the source from which charging current is drawn is increasing directly with e_C. A loop equation beginning at ground would be

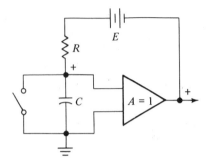

FIGURE 7-16. Simplified bootstrap circuit.

$$Ae_C + E - iR - e_C = 0 \qquad (7\text{-}15)$$

and

$$A = 1, \qquad i = \frac{E}{R},$$

which is a constant current and is the requirement for producing a linear voltage.

In practice, the requirement for a separate supply voltage for the charging circuit is not desirable and can be overcome by employing a large capacitor, which acts as a supply and which replenishes itself from the regular plate supply between sweep cycles. The amplifier is commonly an emitter follower or cathode follower with gain equaling 1. For these reasons, the linearity of the bootstrap sweep circuit is less than perfect but considerably better than that of a normal *RC* circuit.

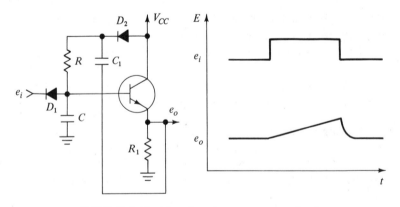

FIGURE 7-17. Transistor bootstrap sweep circuit.

An actual bootstrap circuit is shown in Fig. 7-17. In this circuit, capacitor C_1 replaces the supply voltage, and it should be large compared to the sweep capacitor C. The input signal is derived from a source that has zero potential except when a sweep is desired. In the quiescent state, then, the input to D_1 is clamped to zero volts and T_1 is cut off.

The application of a positive gate at the input diode D_1 cuts it off, and C begins to charge through R and D_2. The voltage on C becomes the input signal to T_1, which is an emitter follower. The output signal is taken across R_1, but it is also fed back through C_1 to maintain a constant potential across R and therefore a constant charging current. When the input gate is removed, the circuit recovers rapidly because C discharges through diode D_1.

Review Questions

1. Explain the meaning of the term *ideal ramp*.
2. State several electronic systems in which sweep generators are used.
3. Define the terms *slope error* and *displacement error*.
4. Define the term *Miller capacitance*.
5. State the meaning of the following terms as related to gas tubes: (a) ignition potential (E_I), (b) extinction potential (E_D), (c) de-ionization current (I_D).
6. What are the major differences between the neon bulb oscillator of Fig. 7-6 and the thyratron oscillator circuit of Fig. 7-8?
7. Explain why the circuit criteria of Eq. (7-4) must be maintained for the neon bulb oscillator to be free running.
8. What are some of the disadvantages of the neon bulb oscillator of Fig. 7-6?
9. State the purpose of resistors R_g and R_2 in the circuit of Fig. 7-8.
10. What is the meaning of the term *bottoming* as related to a pentode tube?
11. Explain how the bootstrap circuit of Fig. 7-17 improves linearity of the output voltage.
12. What effect does E_{BB} voltage have on the frequency of the thyratron sweep circuit illustrated in Fig. 7-8? Explain.
13. What components of the circuit of Fig. 7-10 determine the charge and discharge time constants?
14. What is meant by a voltage waveform having a constant slope over an interval of time?
15. Explain the functional difference between the Miller sweep circuit and the Phantastron sweep circuit.
16. What factors affect the sweep time in the circuit of Fig. 7-15?

17. State the quiescent condition of the circuits in Figs. 7-6, 7-11, 7-14, and 7-15.
18. What is the purpose of D_2 in the circuit of Fig. 7-17?
19. What are the frequency-determining components of the circuit of Fig. 7-8?
20. What factor has the most influence on frequency of a gas tube oscillator, R, C, or V_{BB}? Explain.
21. What precautions should be taken to prevent damage to the transistor in the integrator circuit of Fig. 7-11?

Problems

1. Trace the universal time charge curve $1 - e^{-t/RC}$ on graph paper and solve Example 7-1 for displacement and slope error for time constants of $\frac{1}{2}$ and 2. Compare this result with previous answers and make an appropriate statement concerning linearity of sweep.
2. Consider the circuit of Fig. 7-6 with the following parameters: $R = 50$ kΩ, $C = .02$ μF, $E_D = 54$ V, $E_I = 69$ V, and $t = 100$ μsec.
   Find: (a) E_{BB}
         (b) frequency of oscillation
         (c) $E_{\text{p-p(out)}}$
3. Increase E_{BB} to 800 V in Example 7-2 and solve for the frequency. All other values remain the same. What is implied in this answer?
4. Solve the problem of Example 7-3, neglecting the quantity $I_D R_1$, and compare this answer to the example. What conclusions can be drawn from the results?
5. Draw a schematic diagram of a PNP bootstrap circuit and indicate with dotted lines the charge and discharge paths of the sweep capacitor.
6. Draw a schematic diagram, including the associated waveforms, of an NPN transistor integrator. Indicate with dotted lines the charge and discharge path for C.
7. The thyratron of Fig. 7-8 has the characteristics illustrated by the curve of Fig. 7-7. If the circuit values for the thyratron oscillator are $R_1 = 1$ kΩ, $R_2 = 75$ kΩ, $C = .01$ μF, $E_{BB} = 250$ V, $E_I = 50$ V, $F = 9.5$ kHz, and $I_D = 1$ mA, determine E_C, E_D, and t.
8. Determine the (approximate) maximum current in resistors R_1 and R_2 in Problem 7.
9. What is the output voltage at the end of gate pulse of the circuit illustrated by Fig. 7-10 if the circuit values are 3.3 megΩ, 47 kΩ, .1 μF, 300 V, 200 μsec, and 35 V for R_g, R, C, $B+$, T, and $E_{b(\text{sat})}$, respectively?

10. Draw the resultant output waveform of the circuit of Fig. 7-17 if D_2 shorted.
11. Determine the stated quantities of the circuit of Fig. 7-12(b) if the circuit values are 200 pF, 10^5, 10^5, and $+25$ V for C, A, R, and E, respectively.
    Find: (a) C_{in}
    (b) V_o at $t = .1$ μsec
    (c) τ
    (d) V_o at $t = 1$ μsec
12. Which of the output voltages in Problem 11 would be the more linear? Justify your answer.

Negative Resistance Devices

8

A negative resistance device is one that has, at some part of its voltage-current characteristic, a negative slope. The use of the device in this negative resistance region frequently is well suited to switching circuitry. A well-known device with a negative resistance region is the tetrode. Its characteristics are shown in Fig. 8-1. Note that in the negative resistance region, an increase in plate voltage results in a decrease in plate current. In the tetrode, this action occurs only for relatively low plate voltages, and in most applications the tetrode is operated in the region beyond the negative resistance range. Because of the inherent disadvantages of tubes, tetrodes are not commonly employed in switching applications. Mentioning them here serves only to illustrate the fact that devices other than those we will emphasize exhibit

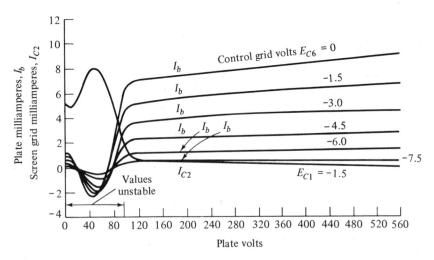

FIGURE 8-1. Plate characteristics of the tetrode.

128

negative resistance. The negative resistance devices we will be mainly concerned with include tunnel diodes and unijunction transistors; PNPN diodes and silicon controlled switches also fall into this category.

8-1 THE TUNNEL DIODE

The tunnel diode is a two-terminal device which has a negative resistance characteristic as shown in Fig. 8-2(b). Fundamentally, the tunnel diode is a heavily doped PN junction diode. The impurity level is so high that when

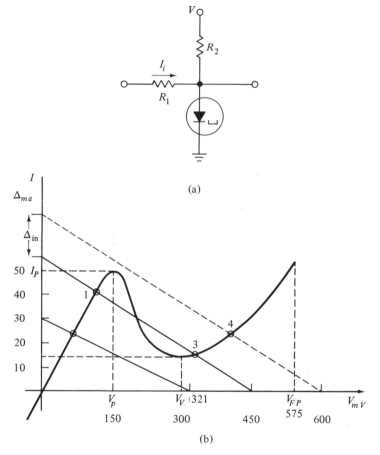

FIGURE 8-2. (a) Tunnel diode switch; (b) characteristics of the tunnel diode.

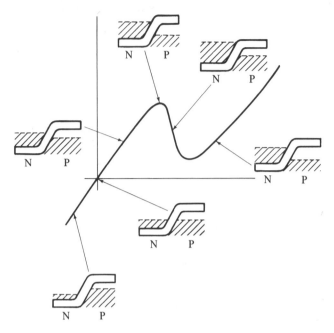

FIGURE 8-3. Energy level diagrams for the tunnel diode.

the junction is formed, the energy level diagram differs from that for a normal junction diode, as shown in Fig. 8-3. The application of a small forward bias to the tunnel diode causes an increase in the energy level of the N material conduction band electrons, and current flows proportionately. Further increasing the forward bias results in increasing diode current until the peak point is reached, where there is maximum availability of conduction band electrons for the valence band holes, as shown in Fig. 8-3. Increasing the forward bias beyond this point results in a reduction of the available electrons for conduction; this trend continues until the valley point is reached, where the least number of charge carriers can cross the junction. At this point, the energy level diagram shows that normal diode action will take place for any further increase in forward bias. The conduction mechanism whereby electrons from the conduction band of the N material combine with the holes in the P material valence band is known as quantum mechanical tunneling, hence *tunnel diode*.

The important parameters of the tunnel diode are the peak point current and voltage, valley point current and voltage, negative conductance, and junction capacitance. In addition, packaging of the diode in a container with leads results in series resistance and inductance that must be considered at high frequencies. Maximum currents and voltages, along with some temperature characteristics, are generally supplied by the manufacturer.

8-2 THE TUNNEL DIODE AS A SWITCH

The phenomenon of the tunneling mechanism, which produces the negative resistance in the tunnel diode, suggests that the device will perform at high frequencies. This effect is true because tunneling is a majority carrier effect with no limitation of minority carrier drift time. There is a great potential for the tunnel diode in high-speed switching applications. When properly biased and triggered, the device can be employed in either a monostable mode or a bistable mode of operation. Furthermore, it can be biased in an unstable region, where it becomes an astable multivibrator.

Simply stated, the ability to employ a tunnel diode in any switching function depends on the selection of the operating point. Figure 8-2 shows a simple tunnel diode switching circuit and the associated characteristic curve. When the circuit is just energized, point 1 (center line) represents the operating point. Application of an input current (ΔI_N) causes the load line to shift, as indicated by the dashed line, and the operating point to change to point 4. Point 4 is located on the positive slope of the curve, representing a stable condition, and the voltage across the diode will remain constant. The circuit remains in this position until the current in the diode is reduced to zero, at which time the original operation point (point 1) is established.

This is a brief discussion of the switching action of tunnel diodes. More detailed analysis of this action will follow when the monostable, bistable, and astable modes of operation are studied.

8-3 THE TUNNEL DIODE ONE-SHOT CIRCUIT

A simple one-shot circuit is shown in Fig. 8-4(a). It consists simply of a potential source, resistor, inductor, and the tunnel diode. Figure 8-4(b) shows the steady state condition for the circuit where the resistance load line is superimposed on the diode characteristic. The requirement for monostable operation is that the load line must intersect the characteristic curve in only one of the positive resistance parts of the curve. Point 1 in Fig. 8-4 is this point.

Circuit operation is quite simple. The application of a positive trigger pulse causes diode current to increase. This current increase follows the curve from point 1 to point 2, the peak current points. If the peak current point is

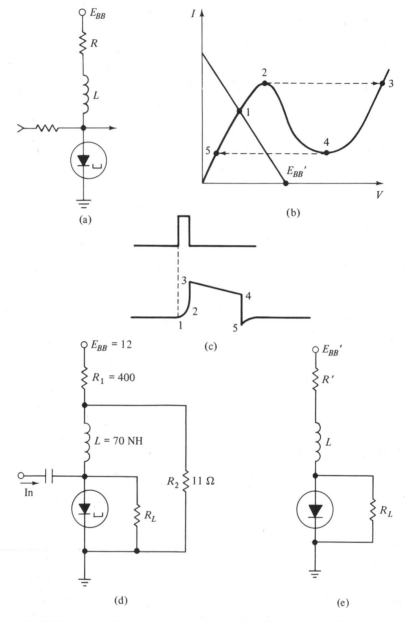

FIGURE 8-4. (a) A tunnel diode one-shot; (b) load-line analysis; (c) waveform of switching action; (d) practical one-shot tunnel diode circuit; (e) simplified circuit.

The Tunnel Diode One-Shot Circuit

not reached, the switching cycle will not occur. When the current reaches point 2 there is an instantaneous jump in voltage across the tunnel diode as the operating point jumps from point 2 to point 3. Since the current does not change instantaneously (it cannot because of the series inductor), this action is permissible. Next, current in the circuit decreases as the operating point moves from position 3 to position 4. The time for this decrease in current to occur is controlled by the slope of the characteristic (a variable resistance) and by the inductor, which combine to form a nonlinear time constant. As seen from Fig. 8-4, this period is the longest part of the cycle and is the major part of the output pulse duration. When point 4 is reached, there is again an instantaneous transition, this time to point 5 (again, there is not an instantaneous change in current). Finally, there is movement of the operating point from point 5 back to the original quiescent position. The actual voltage excursion across the diode represents the output pulse, and Fig. 8-4(c) shows this excursion, with the key points from Fig. 8-4(b) noted.

The input trigger must be of sufficient duration and amplitude to insure that the operating point is driven from the stable position 1 to the peak current point. It is imperative, however, that the trigger duration be shorter than the output signal to prevent interaction. What can occur is that the presence of the trigger signal for too long a period will result in the circuit's seeking a new stable point, because the long trigger will give the appearance of an increased source potential.

The duration of the output pulse consists mainly of that time required for the operating point to move from position 3 to position 4. This time is given by

$$t = \frac{L}{r_2 + R} \ln \left[\frac{I'' + I_a + I_{in}}{I'' + I_v} \right], \qquad (8\text{-}1)$$

where the symbols used are defined in Section 8-4 and Fig. 8-5.

A very important point needs to be emphasized about practical tunnel diode circuits. The circuits of Figs. 8-2(a) and 8-4(a) are not necessarily practical circuits because of the value of E_{BB}. Usually the supply voltage E_{BB} is considered to be several volts (three to ten). However, in tunnel diode circuits the supply voltage is never greater than .75 V. To analyze a functional circuit, as illustrated in Fig. 8-4(d), an equivalent circuit approach is most helpful. The circuit in Fig. 8-4(e) is the Thévenin's equivalent circuit of Fig. 8-4(d). $E_{BB'}$ and R' are the Thévenin's voltage and Thévenin's resistance. Therefore,

$$E_{BB'} = \frac{R_2(E_{BB})}{R_1 + R_2}$$

and R' is

$$R' = \frac{R_1 R_2}{R_1 + R_2}.$$

The following example will help to clarify this point.

Example 8-1 Determine if the circuit of Fig. 8-4(d) is designed to be stable. Assume that the tunnel diode has the characteristics of Fig. 8-2(b). The tunnel diode is a 1N3860 with the following parameters: $I_p = 50$ mA, $I_v = 6$ mA, $V_p = 150$ mV, $V_{fp} = 575$ mV.

Solution: First, simplify the circuit to that of Fig. 8-4(e). Therefore,

$$E_{BB'} = \frac{E_{BB} R_2}{R_1 + R_2} = \frac{(12)(11)}{411}$$

$$= 321 \text{ mV}.$$

$$R' = \frac{R_1 R_2}{R_1 + R_2} = \frac{(11)(400)}{411}$$

$$= 10.7 \ \Omega.$$

The load line is plotted from this equation:

$$E_{BB'} = I_a R' + V_D$$

$$I_{a(\max)} = \frac{E_{BB'}}{R'} = \frac{321 \text{ mV}}{10.7 \ \Omega} \quad \text{when} \quad V_D = 0 \text{ V}, \ I_{a(\max)} = 30 \text{ mA},$$

and

$$E_{BB'} = V_D = 321 \text{ mV} \quad \text{when} \quad I_a = 0.$$

This load line is the lower line of Fig. 8-2(b). The diode is biased in the positive slope. Therefore, a stable condition does exist.

8-4 LINEARIZATION OF THE CHARACTERISTIC CURVE

Calculation dealings with the characteristic curve become quite complex because of the nonlinearity. It is convenient, therefore, to approximate the actual curve with several linear segments. Calculations based on the assumption of linearity yield solutions like Eq. (8-1) which have been found accurate to within approximately 10 percent of the actual values obtained by measurement. Because of this reasonably good agreement and because it further helps to understand those factors affecting circuit operation, we will discuss linearization of the curve.

Figure 8-5 shows (a) a typical curve and (b) the assumed linear approximation of that curve. The important points on the linearized curve are:

Linearization of the Characteristic Curve

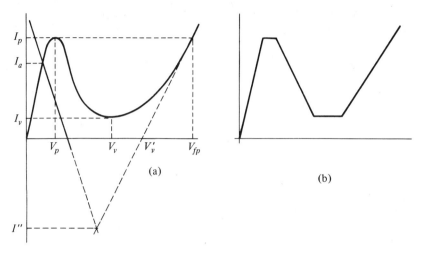

FIGURE 8-5. (a) Tunnel diode characteristics with load-line plot; (b) linearization of the tunnel diode curve.

I_p The peak current point
V_p The peak voltage point
V_{fp} The forward peak point voltage
I_v The valley current point
V_v The valley point voltage
r_1 The dynamic resistance in first positive resistance region
r_2 The dynamic resistance in second positive resistance region
I_a The diode current at the stable point
V'_v Defined as $\dfrac{V_{fp} + V_v}{2}$
I_{in} The input current needed to cause switching.

In the first positive resistance region (low-voltage region), the reciprocal of the slope of the curve is the dynamic resistance, r_1,

$$r_1 \approx \frac{.75 V_p}{I_p}. \tag{8-2}$$

After the curve swings through the negative resistance region, it enters the second positive resistance region. The average (linear) slope of the curve in this region defines the second dynamic resistance, r_2,

$$r_2 \approx \frac{V_{fp} - V'_v}{I_p - I_v}. \tag{8-3}$$

The stable operating point current I_a is either determined graphically or approximated from

$$I_a \approx \frac{E_{BB'}}{R + r_1}. \tag{8-4}$$

8-5 A BISTABLE TUNNEL DIODE CIRCUIT

A circuit which is identical to the monostable circuit can be made into a bistable circuit simply by adjusting the series resistor and the source voltage. The idea is to have the steady state characterized by a condition where the load line intersects both positive resistance regions of the diode curve,

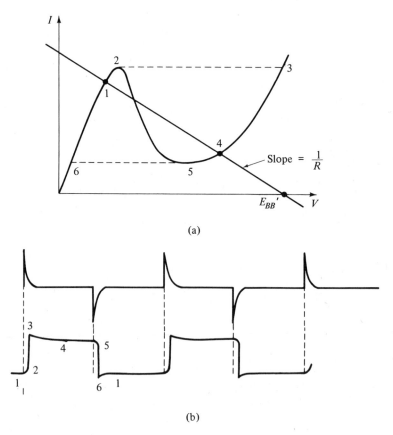

FIGURE 8-6. (a) A tunnel diode flip flop; (b) circuit waveforms.

The Astable Tunnel Diode Circuit

as in Fig. 8-6(a). In this scheme, a positive trigger is required to raise the current from point 1 to the peak current, and then the sharp transition to point 3 occurs, just as for the monostable case. Now, however, as the operating point moves along the curve it finds the second stable position at point 4. Thus the cycle ends, and a positive trigger has resulted in a transition from low voltage to high voltage. To return the diode to point 1, a second trigger, negative this time, is required to move the operating point from position 4 to position 5, and again the fast transition to position 6 takes place, followed by the slower travel back to position 1. Typical waveforms for the cycle are shown in Fig. 8-6(b).

Although this particular bistable circuit is simple, it has obvious disadvantages and is employed only in very simple logic circuits where fast switching is required. The disadvantages of the circuit are that it offers no isolation between input and output, and very small voltage swings are obtainable, generally on the order of 250 mV. It is quite practical, however, to employ tunnel diodes in conjunction with transistors to overcome these disadvantages. Such circuits are commonly called *hybrid circuits* and will be discussed later.

8-6 THE ASTABLE TUNNEL DIODE CIRCUIT

The astable circuit is quite similar in construction to the two previous circuits. Again, the main difference between the astable circuit and others is the choice of source voltage and series resistance. It is obvious from our study to this point that operating points in the negative resistance part of the characteristic curve are *unstable* and that the circuits discussed so far have been stable only at intersections of the load line and *positive resistance parts* of the characteristic curve. It is further apparent that to design a circuit that is astable, one must avoid these stable intersections and seek a load line that crosses the characteristic in the negative resistance region.

Figure 8-7 forms the basis for the analysis. A linearized characteristic is used to make the calculations reasonable. Note that proper placement of the load line so that it intersects the negative resistance part of the curve requires an equivalent source voltage greater than the peak voltage and less than the valley voltage, or

$$V_p < E_{BB'} < V_v. \tag{8-5}$$

The operating point is in continuous motion in this circuit, moving from position 1 to position 2 during one part of the cycle, then jumping rapidly from position 2 to position 3 to begin the second half-cycle. As in the previous cases, transitions from position 2 to 3 and from position 4 to 1 occur almost

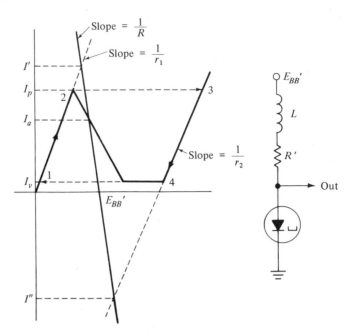

FIGURE 8-7. Linearized characteristic for the astable circuit.

instantaneously and therefore form a negligible part of the total cycle. In other words, the period of the resultant oscillation is

$$T = t_{12} + t_{34} \tag{8-6}$$

and the frequency is

$$f = \frac{1}{T} = \frac{1}{t_{12} + t_{34}}. \tag{8-7}$$

The output signal will be nonsymmetrical because the two positive resistance parts of the characteristic are not equal.

While the operation is between points 1 and 2, the circuit appears as a simple series connection of the source voltage, series resistance (R), inductor (L), and diode resistance (r_1). A Kirchhoff equation for this condition is

$$V_L + ir_1 + iR' = E_{BB'}. \tag{8-8}$$

It can be shown that the solutions of Eq. (8-8), by establishing initial conditions for the periods t_{12} and t_{34}, are:

$$t_{12} = \left(\frac{L}{R' + r_1}\right) \ln \left(\frac{I' - I_v}{I' - I_p}\right) \tag{8-9}$$

Tunnel Diode Hybrid Circuits

$$t_{34} = \left(\frac{L}{R' + r_2}\right) \ln \left(\frac{I_p + I''}{I_v + I''}\right). \tag{8-10}$$

Example 8-2 Consider the astable circuit of Fig. 8-7 with the following circuit values: $E_{BB'} = 200$ mV, $L = 50$ nH, $R' = 2$ Ω, $I_v = 5$ mA, $I_p = 45$ mA, $I' = 55$ mA, $I'' = 50$ mA, $V_p = 150$ mV, $V_v = 300$ mV, $V_{fp} = 550$ mV.

Find: (a) t_{12}
     (b) t_{34}
     (c) frequency

Solution: r_1 and r_2 must be determined. Therefore,

$$r_1 = \frac{.75 V_p}{I_p} = 0.75 \frac{150 \text{ mV}}{45 \text{ mA}} = 2.5 \text{ Ω}$$

$$V'_v = \frac{V_{fp} + V_v}{2} = \frac{550 \text{ mV} + 300 \text{ mV}}{2} = 425 \text{ mV}$$

$$r_2 = \frac{V_{fp} - V'_v}{I_p - I_v} = \frac{550 \text{ mV} - 425 \text{ mV}}{45 \text{ mA} - 5 \text{ mA}}$$

$$= \frac{125 \text{ mV}}{40 \text{ mA}} = 3.125 \text{ Ω}.$$

Now, from Eq. (8-9),

$$t_{12} = \frac{50 \times 10^{-9}}{4.5} \ln \left[\frac{50 \text{ mA}}{10 \text{ mA}}\right]$$

$$= 17.87 \text{ nsec.}$$

And

$$t_{34} = \frac{50 \times 10^{-9}}{5.125} \ln \left[\frac{95}{55}\right]$$

$$= 9.75 \times 10^{-9} \ln (1.727)$$

$$= 5.28 \text{ nsec.}$$

$$f = \frac{1}{t_{12} + t_{34}} = \frac{10^9}{23.15}$$

$$= 43.1 \text{ mHz.}$$

8-7 TUNNEL DIODE HYBRID CIRCUITS

As previously mentioned, the tunnel diode circuit, which is generally the ultimate in simplicity, has several shortcomings. It is worth noting, however, that the advantages of the tunnel diode can be retained and the disadvantages

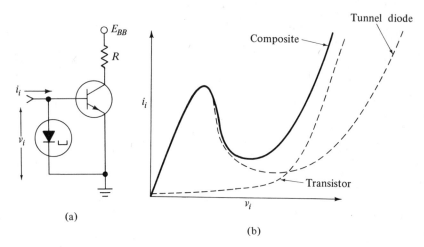

FIGURE 8-8. (a) A basic hybrid circuit; (b) the composite characteristic.

circumvented frequently by constructing a hybrid circuit. The hybrid circuit generally has a tunnel diode connected as part of the input circuit for a transistor stage. In this way, the good response of the tunnel diode is combined with the isolation and large signal output available from the transistor.

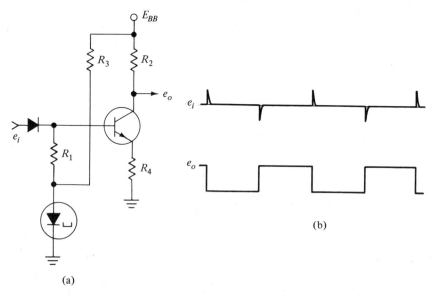

FIGURE 8-9. (a) Simple hybrid flip flop; (b) input-output waveforms.

The Unijunction Transistor

It is interesting to note that simple coupling of a tunnel diode (as in Fig. 8-8) with a transistor results in a modified input characteristic which retains a negative resistance region. In other words, the hybrid circuit of Fig. 8-8(a) has an input characteristic as shown in Fig. 8-8(b), and, therefore, load line analysis as performed in all of the circuits above can be applied to the composite characteristic. It then becomes a simple matter to determine the transistor output based on the division of input current between the tunnel diode and the transistor.

Note from the composite characteristic of Fig. 8-8(b) that operation of the tunnel diode in the high-voltage region will result in high base current to the transistor, whereas low-voltage operation of the tunnel diode yields nearly zero input current to the transistor. A simple flip flop is shown in Fig. 8-9. Positive input triggers switch the diode to the high-voltage state, resulting in increased base current to the transistor. This current, when multiplied by the current gain (β), flows through R_2, and collector voltage drops. Negative triggers are required to return the circuit to its original condition.

8-8 THE UNIJUNCTION TRANSISTOR

The unijunction transistor (UJT), originally called a double-base diode, is constructed from a bar of N-type silicon to which is alloyed a P-type contact, forming a PN junction. Two ohmic contacts from the base bar are designated B_1 and B_2, as shown in Fig. 8-10, which also shows the accepted symbol. The circuit of Fig. 8-10(c) is an equivalent of the UJT when $I_E \leq I_p$. When the equivalent circuit emitter diode conducts, the emitter anode must be more positive than the cathode. The voltage across r_{b_1} is equal to

$$\frac{V_{BB} r_{b_1}}{r_{b_1} + r_{b_2}},$$

and the peak voltage V_p must be a diode drop more positive than r_{b_1} for conduction to occur. The terminology associated with the UJT is:

V_{BB} Interbase potential, the voltage applied between B_1 and B_2
R_{BB} Interbase resistance, the total ohmic resistance of the base (this will vary as a function of emitter current) = $r_{b_1} + r_{b_2}$
I_E, V_E Emitter current and saturation voltage
r_{b_1} Resistance between emitter contact and B_1
r_{b_2} Resistance between emitter contact and B_2
η Intrinsic standoff ratios, $\eta \equiv r_{b_1}/R_{BB} = r_{b_1}/(r_{b_1} + r_{b_2})$
V_p, I_p Peak point voltage and current

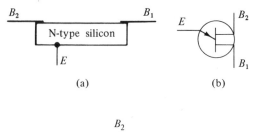

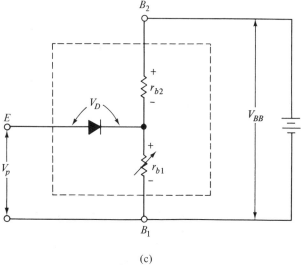

FIGURE 8-10. (a) The unijunction transistor (UJT) construction; (b) schematic symbol; (c) equivalent circuit of UJT, when $I_E \leq I_P$.

V_v, I_v Valley point voltage and current
I_{E_0} Reverse saturation current.

Typical input and output characteristics of the device are shown in Fig. 8-11. The input characteristics are plotted with current on the horizontal axis and are single-valued for any current but can be multiple-valued for a specific voltage. It is interesting to note that the tunnel diode is a voltage controlled negative resistance device and that the UJT, SCR, SCS, and thyristors are all current controlled negative resistance devices. To illustrate all of the data, which are not generally supplied in the characteristics of Fig. 8-11, refer to the sketch of Fig. 8-12.

The cutoff region shown in Fig. 8-12 represents operation of the UJT with the emitter junction reverse biased. With zero emitter voltage applied, a

The Unijunction Transistor

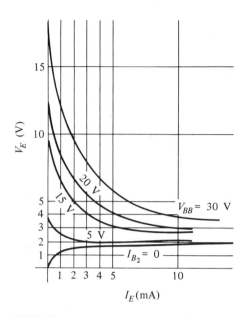

FIGURE 8-11. Input and output characteristics for the UJT.

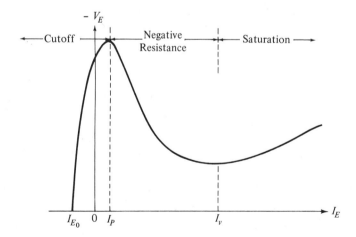

FIGURE 8-12. An entire UJT input curve.

reverse leakage current I_{E_0} flows through the emitter base junction. A decrease in the magnitude of this reverse current is accompanied by an increase in emitter voltage until no current flows at all, and then a small forward current flows until the voltage is increased to the peak voltage V_p. When V_p is reached, the emitter junction is forward biased and holes are injected into the base region. Actually, the transition from reverse bias to forward bias of the emitter junction is a gradual one, and the degree of forward bias is variant over the range of the negative resistance region.

In the negative resistance region, the injection of holes into the base bar results in a decrease of the resistivity of the base bar. As a result of this, an increase in emitter current causes a progressive decrease in resistance between the emitter and base B_1, to which the holes are drawn. The effect of increasing current and decreasing resistance appears as a negative resistance when plotted on a characteristic like that of Fig. 8-12.

The last section of the characteristic curve is the saturation region. It begins at the valley point, that point where no further decrease in base resistivity results from an increase in emitter current. In the saturation region the resistance is low, as indicated by the gentle shape of the characteristic.

The intrinsic standoff ratio, $\eta = r_{b_1}/R_{BB}$, is a parameter which is reasonably constant over a wide temperature range and generally varies between .5 and .75. The peak point voltage is related by the intrinsic standoff ratio to the interbase voltage by

$$V_p = \eta V_{BB} + V_{on}, \qquad (8\text{-}11)$$

where V_{on} is the turn-on voltage of the emitter base junction. ($V_{on} \approx .7$ V typically.) Equation (8-11) shows what is also shown in Fig. 8-11—that is, that variations in interbase potential determine the magnitude of the peak point voltage. Furthermore, since η is relatively insensitive to temperature, the equation also points out that V_p is not adversely affected by temperature, especially when operating at potentials that are considerably larger than V_{on}, which is somewhat temperature sensitive.

The output characteristic curves that accompany the curves of Fig. 8-11 are shown in Fig. 8-13. These would correspond to the collector characteristics of a normal transistor but are actually less significant than the input curves because it is the negative resistance characteristic of the UJT that is important.

Example 8-3 A particular UJT has the following device parameters at 25°C: $\eta = .6$, $R_{BB} = 6.7$ kΩ, $I_{E_0} = 50$ nA, $I_v = 6$ mA, $V_v = 2$ V, $V_D = .5$ V, $I_p = .4$ mA, and $V_{BB} = 25$ V.
Find: (a) r_{b_1}, r_{b_2}
      (b) V_p

A Unijunction Relaxation Oscillator

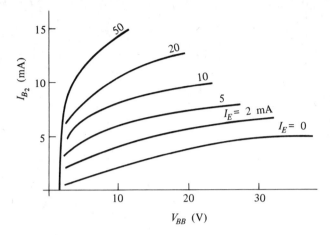

FIGURE 8-13. Output characteristics for the UJT.

(c) Approximate the negative resistance of the diode. (Change of resistance in negative region is ΔR.)

Solution:

$$\eta = \frac{r_{b_1}}{R_{BB}} \quad \text{and} \quad r_{b_1} = (.6)(6.7 \text{ k}\Omega) = 4.02 \text{ k}\Omega.$$

Therefore,

$$r_{b_2} = R_{BB} - r_{b_1} = 6.7 \text{ k}\Omega - 4.02 \text{ k}\Omega$$
$$= 2.68 \text{ k}\Omega$$
$$V_p = V_{BB}\eta + V_D$$
$$= (25).6 + .5 = 15 + .5 = 15.5 \text{ V}.$$

$$\Delta R(\text{negative region}) = R(\text{peak point}) - R(\text{valley point})$$
$$= \frac{V_p}{I_p} - \frac{V_v}{I_v} = 38.42 \text{ K}\Omega.$$

8-9 A UNIJUNCTION RELAXATION OSCILLATOR

A simple and frequently employed circuit which takes advantage of the negative resistance characteristic of the UJT is a relaxation oscillator. The circuit employs a controlling capacitor very much like the gas tube sweep circuit; however, the UJT is a current controlled device, and to understand the circuit operation one must bear this in mind. The basic circuit is shown in Fig. 8-14 along with two possible output waveforms. In order for the circuit

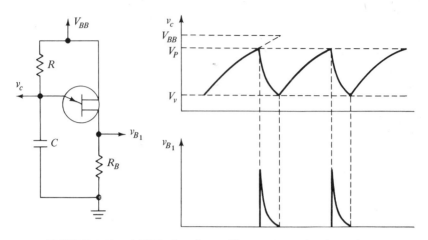

FIGURE 8-14. A UJT relaxation oscillator and associated waveforms.

to function, it is necessary that sufficient emitter current be drawn from the supply to exceed the peak current I_p, but the valley current I_v must not be exceeded. This is simply another way of saying that for oscillation to occur, the load line determined by V_{BB} and R must intersect the negative resistance part of the input characteristics. Assuming these conditions to be met, then, circuit operation consists simply of the charging of C through R until the voltage on C reaches the peak point voltage. When this occurs, the UJT comes out of cutoff. Emitter current increases rapidly as emitter voltage drops. The resultant discharge of the capacitor during the one time of the UJT is accompanied by a sharp output pulse developed across the base resistor. The abrupt loss of capacitor charge, to the point where the valley voltage is reached, results in the cutoff of the UJT. The cycle then repeats. The charge time for the capacitor, while the UJT is off, is found from the familiar RC equation when the capacitor is charging from one potential toward another:

$$t = RC \ln \frac{V_{BB} - V_v}{V_{BB} - V_p}. \tag{8-12}$$

This equation is approximately the same as

$$t = RC \ln \frac{1}{1 - \eta}, \tag{8-13}$$

since

A Unijunction Relaxation Oscillator

$$V_p = \eta V_{BB} + V_D \qquad (8\text{-}14)$$

and the ratios of V_D/V_{BB} and V_v/V_{BB} are reasonably small.

The student should be careful to note that V_{BB} in the circuit described is identical to the supply voltage—a condition that is not true if a base resistor is inserted in the B_2 lead. If such a resistor were employed, then the supply voltage would be greater than V_{BB} (the interbase potential).

Example 8-4 Solve the circuit of Fig. 8-14 for the stated quantities; the circuit values are: $R = 10 \text{ k}\Omega$, $F = 4 \text{ kHz}$, $R_B = 10 \text{ }\Omega$, $V_{BB} = 14 \text{ V}$, $V_v = 2 \text{ V}$, $R_{BB} = 9 \text{ k}\Omega$, $\eta = .7$, $V_D = .5 \text{ V}$.

Find: (a) V_p
(b) C
(c) r_{b_1}
(d) r_{b_2}
(e) $I_{B_1(\max)}$

Solution: V_p is solved by Eq. 8-14.

$$\begin{aligned} V_p &= \eta V_{BB} + V_D \\ &= (.7)14 + .5 \\ &= 9.8 + .5 \\ &= 10.3 \text{ V}. \end{aligned}$$

C can be solved for from Eq. 8-12:

$$t = RC \ln \frac{V_{BB} - V_v}{V_{BB} - V_p}, \quad \text{where} \quad t = \frac{1}{f}$$

$$250 \times 10^{-6} = 10^4 C \ln \left[\frac{14-2}{14-10.3}\right]$$

$$= .0213 \text{ }\mu\text{F}.$$

Recall that

$$\eta = \frac{r_{b_1}}{R_{BB}}$$

and

$$r_{b_1} = \eta R_{BB} = (0.7)9 \text{ k}\Omega = 6.3 \text{ k}\Omega$$
$$r_{b_2} = R_{BB} - r_{b_1}$$
$$= 9 \text{ k}\Omega - 6.3 \text{ k}\Omega = 2.7 \text{ k}\Omega$$

Also,

$$\begin{aligned} I_{B_1(\max)} &= \frac{V_{B_1(\max)}}{R_B} \\ &= \frac{V_p - V_v}{R_B} = \frac{11.8 - 2 \text{ V}}{10} \\ &= 980 \text{ mA}. \end{aligned}$$

The value for R_B should be selected so that the maximum emitter current is not exceeded. The magnitude of the output pulse is V_{B_1}(9.8 V).

8-10 NEGATIVE RESISTANCE DEVICES IN GENERAL

It has been demonstrated in the preceding pages that the negative resistance characteristic of the tunnel diode and the unijunction transistor make possible the design of circuits in any of the three basic modes—that is astable, bistable, and monostable. In general, for any negative resistance device, the type of circuit operation is simply dependent upon the placement of the load line on the device characteristic. If the load line intersects only the negative resistance part of the characteristic, then the circuit will be astable. A monostable circuit will be one where the load line intersects the characteristic in a stable region only once, whereas the bistable circuit is developed by placing the load line so it crosses the characteristic curve at two stable points. With these thoughts in mind, the student is urged to conceive a one-shot circuit employing a UJT and also a UJT flip flop.

Other devices exist that exhibit the negative resistance characteristic, and the comments above apply to these units also. Among the more recently developed and commercially available devices are the four-layer diode, the backward diode, and the silicon controlled switch.

Review Questions

1. Define the term *negative resistance*.
2. State the conditions that must exist in a PN junction for the phenomenon of tunneling to occur.
3. Explain the meaning of the parameter peak point current and of the peak point voltage that are associated with tunnel diodes.
4. Briefly describe the switching actions that occur when a tunnel diode is operating in a switching mode.
5. Define the term *eta* (η) as it relates to a unijunction device.
6. Explain the basic difference between a voltage controlled negative resistance device and a current controlled negative resistance device.
7. What is the meaning of *stable operating point*?
8. What are the requirements placed upon the trigger pulse of a monostable tunnel diode circuit?
9. Explain the meaning of the term *interbase resistance* (R_{BB}).
10. What occurs in the emitter of a UJT circuit when the voltage (V_p) is slightly exceeded in value? Explain.

11. What is the relative value of the resistance of a UJT in its saturation region compared to the resistance in a cutoff region? Explain.
12. Why is r_{b_1} in Fig. 8-10(c) shown as a variable resistor?
13. Explain the meaning of the circuit parameters $E_{BB'}$ and R' as illustrated in Fig. 8-4(e).
14. State a major advantage and a major disadvantage of a tunnel diode switch.
15. What is the meaning of the *unstable* region of a tunnel diode characteristic curve?
16. State at least three other types of electronic device that exhibit negative resistance characteristics.
17. What is the major factor that characterizes the tunnel diode as a high-speed device?
18. What influence does the interbase potential have on the frequency of a UJT relaxation oscillator? (See Eq. 8-13.)
19. What type of vacuum tube displays a negative resistance characteristic? Explain this action.
20. Is it possible for a tunnel diode astable multivibrator to be symmetrical? Explain.

Problems

1. Refer to the circuit of Fig. 8-4(d) and the graph of Fig. 8-2(b) and determine if the tunnel diode is biased for astable, monostable, or bistable operations. The circuit values are: $R_1 = 500 \, \Omega$, $R_2 = 15 \, \Omega$, $E_{BB} = 15 \, V$.
2. What change in circuit parameters would be needed to convert the circuit of Problem 1 to the other two modes of operation? Show equivalent circuits and values.
3. Determine the following quantities of the circuit of Fig. 8-7 if the circuit and device value sare: $E_{BB} = 250 \, mV$, $L = 62 \, nH$, $r_1 = 3.1 \, \Omega$, $r_2 = 4.2 \, \Omega$, $V_{fp} = 525 \, mV$, $V_p = 150 \, mV$, $V_v = 275 \, mV$, $R' = 2.2 \, \Omega$, $I' = 46 \, mA$, $I'' = 41 \, mA$.
   Find: (a) V'_v
   (b) I_v
   (c) I_p
4. Determine the frequency of oscillation for the circuit in Problem 3.
5. Determine $E_{BB'}$ for the circuit of Problem 1.
6. What change would occur in circuit operation of Fig. 8-4(d) if E_{BB} and R_1 were 6 V and 110 Ω, respectively? All other values remain as indicated. (See Example 8-1.)

7. The device of Fig. 8-11 has the following parameters: $r_{b_2} = 3$ kΩ, $V_v = 2$ V, $V_p = 20$ V, $V_D = .5$ V, $I_p = .3$ mA, $I_v = 4$ mA, $V_{BB} = 27$ V.
   Find: (a) η
   (b) r_{b_1}
   (c) R_{BB}
   (d) ΔR

8. Solve the circuit of Fig. 8-14 for the stated quantities if the circuit and device values are: $C = .01$ μF, $R_B = 15$ Ω, $V_{BB} = 20$ V, $\eta = .65$, $V_v = 1.5$ V, $V_D = .6$ V, $F = 10$ kHz, $R_{BB} = 12$ kΩ.
   Find: (a) R
   (b) r_{b_1}
   (c) r_{b_2}
   (d) $V_{b_1(\max)}$

9. Place a 2 kΩ resistor in series with B_2 of the UJT in Fig. 8-14 and solve for V_p and the frequency. Use the value of R obtained in Problem 8. All other values remain constant.

10. Did the 2 kΩ resistor, when placed in the circuit of Problem 9, alter the oscillator frequency significantly? Justify your answer.

11. Resolve the problem of Example 8-4 if the only circuit changes are: 6 kHz, .75, and 20 V for f, η, and V_{BB}, respectively.
    Find: (a) V_p
    (b) C
    (c) r_{b_1}
    (d) r_{b_2}
    (e) $I_{B_1(\max)}$

Index

Active region, 42, 47, 48
Alpha current gain, 36
Amplifiers, transistor, 35
  class A, 35
  common base, 36
  common emitter, 37
  linear operation, 35
Astable multivibrator, 85
Avalanche diode, 20

Basic function, 1
Beta current gain, 37
Bistable multivibrator, 70
Black box analysis, 40
Blocking oscillator, 90
  astable, 99
    synchronization, 102
  monostable, 96
    synchronization, 99
Bootstrap sweep circuit, 123
Bottoming (see Switches)

Clamping circuits, 25
  dc level, 27
  diode clamper, 25
  grid clamp, 49
  nonsymmetrical square, 29
    delay clamp, 31
    ground clamp, 31
    wave clamp, 29
  steady state condition, 27
Clipping circuits, 22
  diode biased clipper, 24
  half-wave rectifier, 22
  ideal diode, 25
  simple diode circuit, 22
  triode clipper, 51
Common base amplifier (see Amplifier, transistor)
Common emitter amplifier (see Amplifier, transistor)
Constant current source, 38

dc restorer, 25
Differentiating circuit, 7
  high-pass filter, 9
  lower cutoff frequency, 9
  RC differentiator, 7, 8, 24
  RL differentiator, 9
Diode response time, 21
  junction capacitance, 22, 25
  storage delay time, 21, 47
  switching time, 21
  transition time, 21
Diode, semiconductor, 16
  Avalanche, 20
  characteristic curve, 19
  equation, 18
  germanium, 19, 21
  PN junction theory, 16
    depletion region, 16

Diode (cont.):
  forward biasing, 17
  reverse biasing, 17, 19
  saturation current, 18
  silicon, 19, 21
  varactor diode, 18
  Zener, 20
Diode steering, 74

Eccles-Jordan multivibrator (see Multivibrator)

Flip flop, 70
Free-running multivibrator, 85

Gain-bandwidth, 47
Gates, 57
  AND, 57
  bidirectional, 66
  NAND, 56
  nonideal conditions, 68
  NOR, 56
  NOT, 59
  OR, 58
  transmission, 62
  unidirectional, 63

High-pass filter, 9
Hybrid circuits, 137, 139
Hybrid parameters (see Transistor)

Integrated circuit, 120
Integrating circuits, 10
  definition, 11
  low-pass filter, 13
  Miller integrator, 117
  RC integrator, 10
  RL integrator, 12
  transistor integrator, 116
  triode integrator, 115
  upper cutoff frequency, 13
Interbase resistance (see Unijunction transistor)

Logic, basic principles, 54
  DTL, DL, 56
  negative logic, 62
  RTL, DCTL, TTL, 62
Lower cutoff frequency, 9
Low-pass filter, 13

Miller capacity, 118
Miller effect, 118
Miller integrator, 117
Multivibrators, 70
  astable (free running), 85
    nonsymmetrical, 86
    quasistable state, 85
  bistable (flip flop), 70
  cathode coupled bistable, 75
  Eccles-Jordan, 70

Index

Multivibrators (cont.):
  monostable (one shot), 79
    emitter coupled, 83
    equivalent circuit, 79
    triode (one shot), 81
  nonsaturating, 75
  saturating, 74
  Schmitt trigger, 75
    hysteresis, 77
    squaring circuit, 77

Negative resistance, 128, 148

One shot multivibrator (see Multivibrators)

Peak point current (see Tunnel diode)
Pedestal, elimination, 66
Pentode, 38
  plate characteristic, 38, 51
Phantastron sweep, 122
Phase shift, 10
PNPN diodes, 129
Pulse transformer, 90

Ramp function, 106
RC series circuit, 2, 3
  voltage equation, 3
Relaxation oscillator, 90, 145
Rise time, definition, 45 (see also Switches)
RL series circuit, 3, 4
  current equation, 4

Saturation current, 18
Sawtooth voltage, 106
Schmitt trigger, 75
Single-energy storage circuits (see RL, RC series circuits)
Sweep generators (see Time base generators)
Switches, 42
  transistor, 42
    cutoff, 42
    delay time, 45
    fall time, 47
    "forced β," 45
    response of switch, 45
    rise time, 46
    saturation current, 43
    saturation resistance, 43
  tunnel diode, 131
  vacuum tube, 48
    bottoming, 51
    pentode, 51
    triode, 48
Synchronization, 72

Tertiary winding, 90
Tetrode characteristics, 128
Thévenins equivalent circuit, 61
Time base generators, 106
  bootstrap sweep, 123
  displacement error, 109
  gas tube, 110
    extinction potential, 110
    ionization potential, 111

Time base generators (cont.):
  Miller integrator, 117
    pentode sweep, 121
  Phantastron sweep, 122
  ramp function, 106
  sweep linearity, 107
  slope error, 108
  thyratron sweep, 112
    ionization current, 113
    synchronization, 114
  transistor integrator, 116
  triode integrator, 115
Time constant, 4
  definition, 5
  end of transient period, 6
  universal graph, 5
Transformer, pulse, 90
  equivalent circuit, 94
  ideal, 90
  nonideal, 94
  phase shift, 91
  tertiary winding, 90, 98
  turns ratio, 91
Transistor, 38
  collector characteristics, 38
  h parameters, 40
  leakage current, 39
Triggering, 72
  blocking oscillator, 99, 102
  cathode, 74
  common collector, 73
  common emitter, 72
  delay line, 103
  diode steering, 74
  grid, 74
  plate, 74
  speed-up capacitors, 73
Triode switch, 48
Tunnel diode, 128
  astable circuit, 137
  bistable circuit, 136
  characteristics, 129, 135
  fundamental theory, 129
  hybrid circuits, 137, 139
  monostable circuit, 132
    circuit requirements, 131
    equivalent circuit, 133
    load line analysis, 133
    operating point, 131, 133
    practical circuit, 133

UJT (see Unijunction transistor)
Unijunction transistor, 141
  double-base diode, 141
  equivalent circuit, 142
  intrinsic standoff ratio, 141, 144
  relaxation oscillator, 145
  UJT terminology, 141
Upper cutoff frequency, 13

Valley point current (see Tunnel diode)
Varactor diode, 18

Zener diode, 20